Leckie✕Leckie
Scotland's leading educational publishers

Success guides

HIGHER
Geography

✕ Dr Bill Dick ✕

Contents

Introduction

The Physical Environment

The Human Environment

Environmental Interactions

Introduction

Aims of this Success Guide

This *Success Guide* will not cover all of the topics in the Higher syllabus, or give you as much detail as your class texts. However, when you are revising for the exam, it will act as a useful guide to the main points examiners look for in answers. You do not have to remember all of the points covered in each chapter, but you should try to gain a basic knowledge of the topics covered.

The book contains a series of **Quick Tests** and **Top Tips** and, at the end of each chapter, sample answers are given to a range of questions from examinations. These answers have been marked and comments on the quality of the answers are also provided.

When preparing for the exam you will need to cover the Physical and Human Environments and any two Geographical Interactions. The latter will have already have been decided for you by your teacher. Your teacher may be able to help you predict possible questions/topics in the examinations you will sit. This is done on the basis of topics which have been examined in previous examinations. However, this kind of prediction is not a perfect science and should be followed with caution.

Structure of the exam

In 2005 the format of the Higher Geography exams was changed.

There are two Question Papers, each marked out of 100.

The time allocation for Paper 1 is 1 hour 30 minutes.

The time allocation for Paper 2 is 1 hour 15 minutes.

Paper 1 consists of three sections, A, B and C.

Section A contains four compulsory questions – two topics on the Physical Environment and two on the Human Environment.

Section B has questions on the two Physical Environment topics not examined in section A of which you have to answer one.

Section C has questions on the two Human Enviroment topics not examined in section A of which you have to answer one.

Paper 2 Environmental Interactions consists of six questions.

You must answer any two of these questions.

Each question in this paper is marked out of 50.

Examination advice

- When revising you can use this guide together with past examination questions and revise topic by topic, e.g. Urban – settlement zones or Population – migration factors.
- Use the selected sample question and answer sections to revise. Try the questions and check your answers against the marked answer provided.
- Remember if you mark your answers, award a full mark for every correct statement or appropriate example.

- For the actual examination, prepare your notes into sections.
- Try to work out a schedule for studying with a programme which includes those sections of the syllabus you intend to study.
- If there is a certain amount of predictability in the topics being asked each year, go through past papers to review topics previously asked. This might help you to prioritise your study topics.
- Check your knowledge of the topics from time to time using past paper questions.
- Organise your notes into checklists and revision cards.
- Practise drawing diagrams which may be included in your answers, for example, corries or pyramidal peaks.
- Try to avoid leaving your studying to a day or two before the examination.
- Also try to avoid cramming your studies on the night before the examination, especially staying up late to study.
- Make sure you know the examination timetable, noting the dates and times of your examinations.
- Give yourself plenty of time by arriving early for the examination, well equipped with pens, pencils, rubbers, etc.
- Read the instructions carefully. If asked to describe and explain make sure that you refer to both in your answer.
- If you are asked for a named country or city, make sure you include details of any case study you have covered.
- Avoid vague answers when asked for detail. Avoid general terms such as 'dry soils' or 'fertile soils' if you can give more detailed information, e.g. 'deep and well-drained soils' or 'rich in nutrients'.
- If you are given data in the form of maps, diagrams and tables in the question, make sure you use this information in your answer to support any points of view you give.
- If describing climates, give climate figures.
- Be guided by the number of marks for a question as to the length of your answer.
- Do not spend too much time on any particular answer, thus leaving yourself short of time to finish the paper.
- Try to time yourself during the examination for each question. Make sure that you leave yourself sufficient time to answer all of the questions.
- If you have any time left in the exam, use it to go back over your answers to see if you can add anything to what you have written by way of additional text, including more examples or diagrams which you may have omitted. This is especially helpful in Ordnance Survey based questions.
- Make sure that you have read the instructions on the question carefully and that you have avoided needless errors such as answering the wrong sections, or failing to explain when asked to, or perhaps omitting to refer to a named area or case study.
- One technique which you might find helpful, especially when answering long questions worth 10 or more marks is to 'brainstorm' yourself for possible points for your answer. You can write these down in a list at the start of your answer.
- As you go through your answer you can double check with your list to ensure that you have put as much into your answer as you can. By doing this you avoid coming out of the exam and being annoyed that you forgot to mention an important point.

Common errors

Markers of the external examination often remark on errors which occur frequently in candidates' answers. These include the following:

Lack of sufficient detail

- This often occurs in Higher case study answers, especially in 10 to 18 mark questions.
- Many candidates fail to provide sufficient detail in answers, often by omitting reference to specific examples, or not elaborating or developing points made in their answer.
- As noted earlier, a good guide to the amount of detail required is the number of marks given for the question. If, for example, the question is worth 10 marks, then you should make at least six valid points for a pass.

Listing

If you give a simple list of points rather than fuller statements in your answer you may lose marks; for example, in a 4 mark question, you will obtain only 1 mark for a list.

Bullet points

The same rule applies to a simple list of bullet points. However, if you give bullet points with some detailed explanation you could achieve full marks.

Irrelevant answers

You must read the question instructions carefully so as to avoid giving answers which are irrelevant to the question. For example, if asked to explain, and you simply describe, you will not score marks. If asked for a named example and you do not provide one, you will forfeit marks.

Repetition

You should be careful not to repeat points already made in your answer. These will not gain any further marks. You may feel that you have written a long answer, but it may contain the same basic information repeated again and again. Unfortunately these statements will be ignored by the marker.

Reversals

Occasionally questions involve opposites. For example, an answer to such a question might say: 'Death rates are high in developing countries due to poor health care' and then go on to say 'Death rates are low in developed countries due to good health care'. Avoid doing this. You are simply stating the reverse of the first statement. A better second statement might be 'High standards of hygiene, health and education in developed countries have helped to bring about low death rates'.

Atmosphere

The study of **Atmosphere** forms the first part of the Physical Environment element of the Higher Geography course. This involves an understanding of a wide range of concepts and theories relating to variations in world temperature and rainfall patterns and their effects. For the purposes of the examination, this can be reduced to the study of a specific number of aspects of atmosphere and climate.

Global scale

How to describe and explain the earth's heat budget

Why does the earth receive only about 46% of the solar insolation which leaves the sun? This is often asked in the form of **'explain the earth's heat budget'**.

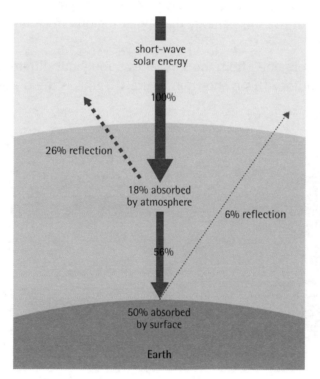

FIGURE 1.1: EARTH'S HEAT BUDGET

- Figure 1.1 shows a summary of the energy which the earth receives from the sun and how it is distributed.
- The earth is heated by solar rays from the sun.
- Some of this heat is absorbed directly by land and water, some is interrupted by clouds and dust in the atmosphere and some is reflected back into space before even reaching the earth's surface.
- For every 100 units of energy which the earth receives, 31 units are reflected back into space (from clouds (17), gases and dust (8) and from the earth's surface (6)).
- This reflected energy is called the **earth's albedo**.
- A further 23 units are absorbed by clouds, water vapour, dust and various gases.
- The remaining 46 units are absorbed by the earth's surface land and water.

The Physical Environment

What factors affect the amount of sunlight reflected from the earth's surface?

If asked to discuss this, refer to:

- Energy absorbed by the earth causes the earth's temperature to rise; some energy bounces back from the surface into the atmosphere where it is absorbed by clouds and gases.
- These clouds and gases are heated and they in turn produce energy, some of which is returned to the earth's surface. This once again heats the earth's surface temperatures.
- Most of the energy which heats the atmosphere actually comes from the earth's surface.
- This happens through a process called **conduction**, which is energy rising from the earth's surface.

Top Tip

Remember that reflection rates vary – ice, snow and water have high rates while forest cover has a low rate.

Why are lower latitudes warmer than higher latitudes?

This asks, in effect, why equatorial areas are very warm while polar areas are very cold. This is due to lower latitudes receiving more of the sun's energy than polar regions. You can use a diagram if asked to explain this.

The amount of heat or energy obtained from the sun varies throughout different parts of the earth. This is due mainly to the effect of latitude. This is shown in Figure 1.2 (this is the diagram which you could use in an answer).

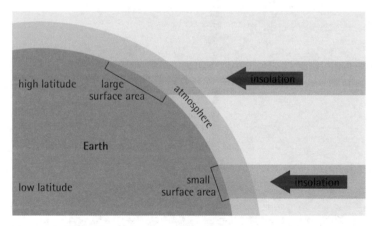

FIGURE 1.2: VARIATION IN EARTH'S INSOLATION

When discussing the effect of latitude on the distribution of solar energy, refer to the following:

- Because the earth is a sphere and the sun's rays strike the areas around the centre of the earth at right angles, places at the equator or between the tropics are always hotter than places at higher latitudes.
- At higher latitudes the rays strike the surface at a wider angle. This means places nearer the poles receive less heat from the sun.
- The rays striking the surface at the equator heat up a smaller surface area than the same amount of rays in areas closer to the poles.

Why is there a net gain of heat from the sun towards the tropics and a net loss of heat towards the poles?

This is due to the earth's movement around the sun.

- The earth revolves around the sun. Therefore the sun is overhead at the equator and each of the tropics at different times of the year.
- At the poles there are alternately six months of light and six months of darkness.

What is meant by 'Global Transfer of Energy'?

Global transfer of energy is due to the following:

- Areas north and south of latitude 38 degrees receive **less solar energy** than areas between latitudes 38 degrees N and 38 degrees S.
- In the higher latitudes more heat is **given out** from the surface than is absorbed.
- Nearer the tropics more energy is **absorbed by than emitted from** the surface.
- Consequently there is a **deficit** in solar energy north and south of latitudes 38 degrees and a **surplus** of solar energy in areas between 38 degrees N and 38 degrees S.
- If this situation remained static, areas near the tropics would become hotter while those further north and south would become colder.
- The fact that this does not happen is due to the transfer of energy from areas of surplus to areas of deficit.
- This transfer of energy is known as atmospheric circulation.

How does atmospheric circulation work?

Atmospheric circulation happens because:

- At the equator the energy at the surface of the earth heats the air immediately above it. This air expands, becomes less dense and rises to a higher altitude creating a zone of low pressure.
- At the higher altitude the temperature is colder and therefore the air cools, becomes more dense and begins to fall. Differences in pressure between the surface and upper atmosphere creates a **wind**.
- Cold falling air moving towards the poles creates high pressure zones around 30 degrees N and 30 degrees S. This circulation of air forms cells both north and south of the equator called **Hadley cells**.
- In addition to the Hadley cells above the tropics, cells exist above each of the poles due to cold air becoming dense and falling to create a high pressure zone.
- These are called **Polar cells**.
- Some of this air moves from the high pressure area towards lower latitudes due to the movement of the earth.
- In these lower latitudes the air is heated and begins to rise into higher altitudes where it is cooled, creating a zone of low pressure.
- Thus a circulation pattern of air occurs at the poles similar to that above the tropics.
- A third cell, termed a **Ferrel cell**, forms due to the temperature differences between the first two cells at the tropics and at the poles.
- Warm air from the Hadley cell at the tropics feeds into the higher latitudes while colder air from the polar cells feeds into the lower latitudes.
- This leads to the transfer of energy from the warmer lower latitudes to the higher colder altitudes and transfer of colder air from the colder higher latitudes to the warmer lower latitudes.

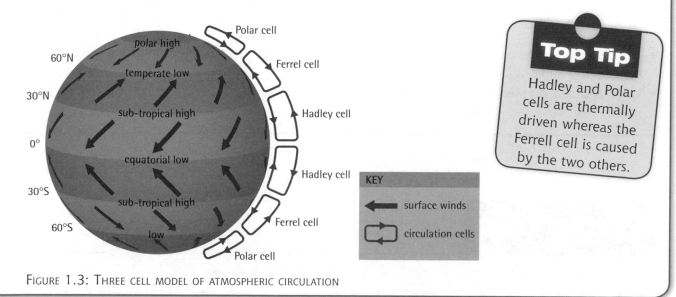

FIGURE 1.3: THREE CELL MODEL OF ATMOSPHERIC CIRCULATION

Top Tip

Hadley and Polar cells are thermally driven whereas the Ferrell cell is caused by the two others.

Describe the pattern of atmospheric circulation and global winds.

Pressure and wind patterns

A simple pattern of distribution of pressure belts is shown in Figure 1.4.

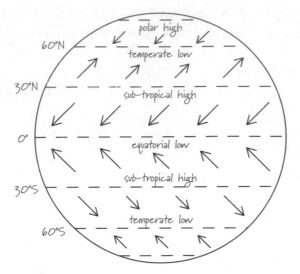

FIGURE 1.4: SIMPLE PATTERN OF DISTRIBUTION OF PRESSURE BELTS

This pressure belt pattern changes throughout the year due to several factors including:

- As air is heated it expands and rises, leaving the pressure near the surface low since there is less air to create pressure. As the air rises it cools and becomes more dense and falls.
- In areas where the air is cooler it is more dense and this creates an area of high pressure.
- Areas north and south of the equator are relatively cooler and are areas of high pressure.
- The air at the poles is also colder and denser and these form areas of high pressure.
- Air blows from high to low pressure creating winds.
- Air will blow outwards from the high sub-tropical areas towards the equator and towards the relatively lower pressure areas between the sub tropics and the poles.

Explain the pattern of atmospheric circulation and global winds.

Reasons include:

- The fact that the **position of the sun** changes during the seasons due to the earth revolving around the sun. This affects the position of the pressure belts which change.
- In December, when the sun moves over the Tropic of Capricorn, the pressure belts move further southwards.
- The **rotation** of the earth also affects pressure belts. This movement tends to deflect air from the poles towards the equator.
- **Land and sea masses** also affect pressure and wind patterns. The rate of heating and cooling varies greatly over land and sea areas.

Quick Test

Referring to Hadley and Polar cells, explain how the Ferrel cell is formed.

- The flow of wind throughout the earth is also affected by **landscape features** such as mountains which deflect winds from their path.
- There are large belts of fast moving winds transversing the globe at high altitudes of between 10 000 and 12 000 metres.
- There are streams of very fast moving air known as **jet streams** within these waves.

What are the main wind patterns throughout the earth?

Main features include:

- The fact that there are basically two main wind belts in each hemisphere, namely the **trade winds** and the **westerlies**.
- Trade winds are found **between latitudes 30 N and 30 S** and are caused by the movement of air from the high pressure sub-tropical zones towards the low pressure zone at the equator.
- The westerly winds flow pole wards **out from the high sub-tropical pressure areas** towards the **middle latitude areas** in the northern and southern hemispheres.
- There are smaller belts of winds flowing outwards from the poles in an easterly direction in both hemispheres which occur only in winter in the northern hemisphere.
- In the southern hemisphere the pattern is not confined to any particular season.
- Areas further inland are less influenced by these winds.

How to describe the general pattern of ocean currents on a world map

- 71% of the earth's surface is covered by water and 29% of the surface is land.
- This has an important influence on the transfer of energy since water is a much more efficient store of heat than land.
- The oceans warm more slowly than land but are heated to a greater depth. Heat is redistributed due to the flow of ocean currents.
- Because the waters nearer the equator receive more heat than those near the poles, warm water flows **outwards** from the equatorial regions towards higher latitudes.
- In turn, colder water from the poles flows **towards warmer regions** creating a circulatory system.
- The flow is disrupted and distorted by the effect of the **earth's rotation** and the distribution pattern of the **world's land masses** creating the pattern of ocean currents which exist at present.

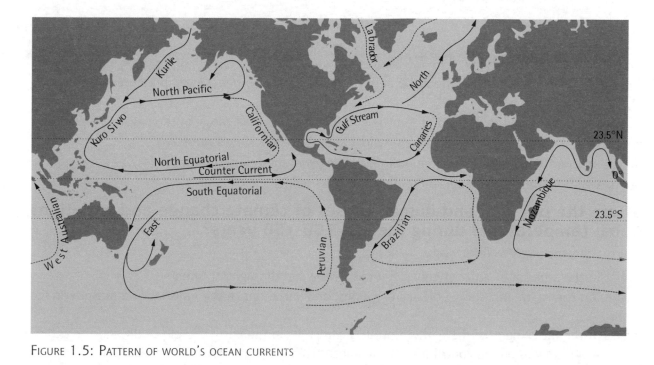

FIGURE 1.5: PATTERN OF WORLD'S OCEAN CURRENTS

How to explain this pattern

When answering, refer to the following:

- The pattern of the world's ocean currents is closely linked with the distribution of the world's main pressure belts and wind patterns.
- The presence of continental land masses distorts the flow of the currents producing the pattern shown in Figure 1.5.
- Winds blowing over these currents would assist in the flow of warm water to cooler areas and cooler water to warmer areas.
- Due to the earth's rotation, winds in the northern hemisphere are deflected to the **right** and in the southern hemisphere to the **left**, helping to create the pattern of currents shown in Figure 1.5.
- If the current is warm or cold and the wind is onshore or offshore, this has a vital effect on climatic conditions on the land masses.
- The pattern of ocean currents has a great influence on the temperature patterns throughout the world over the different seasons.

How have global temperatures varied during the last 100 years?

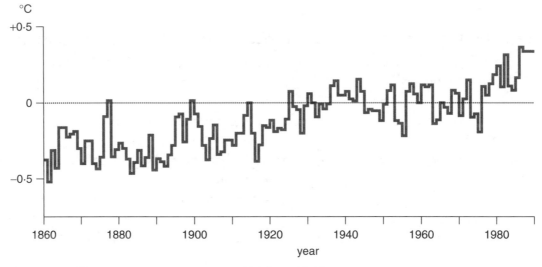

FIGURE 1.6: TEMPERATURE VARIATIONS FROM 1860 TO 1990

When describing the variations on this graph:

- Refer to the general trend throughout the period, mentioning whether the temperature is generally rising or falling.
- Describe the highest and lowest points on the graph and the overall range in temperature change.
- State the periods when temperatures were below and above the stated norm.
- Describe any periods when there were dramatic changes in the temperature.
- Refer to actual figures from the graph to illustrate your points.

What are the physical and human causes of climate change and variations in global temperatures during the last 100-150 years?

Physical reasons include the following:

- Changes in the amount of solar energy given out by the sun throughout time.
- Activities on the earth, including volcanic eruptions, and variations in the amount of atmospheric gases present.
- Changes in the movement of the earth in orbit. Slight shifts in the earth's angle of tilt and the orbit pattern around the sun have contributed to significant changes in the temperature pattern.

- Gases given off from rotting vegetation in tundra areas have affected global temperatures.
- Sunspots also increase temperatures.

Human reasons include the following:

- The wide-scale burning of fossil fuels and forested areas throughout the world has released various chemicals into the air including sulphur dioxide, carbon monoxide and carbon dioxide.
- Trees which gave out valuable oxygen to the air have been cut down and removed in great quantities.
- Increasing industrialisation releasing air pollution from chimneys and factories.
- The increasing release of pollutants from traffic, rubbish dumps or other similar sources, e.g. **CO₂, CFCs and NO₂**. These gases are carbon dixide, fluorocarbons and nitrous oxide and are largely responsible for the **'Greenhouse Effect'**.
- Increase in gases such as methane from large herds of livestock, particularly cattle.
- Releases of CFCs (chlorofluorocarbons) from, for example, aerosols and refrigerants, cause increases in global temperatures.
- The testing of atom bombs releasing radioactive material may have affected the earth's atmospheric conditions.

A number of people believe that many of the human reasons listed have contributed to a process known as **Global Warming**. In effect, temperatures throughout the world are thought to have risen slightly in recent years.

What effect has global warming had on the earth?

Main impacts may include:

- Melting ice caps at the poles.
- A rise in sea levels around coastlines.
- Changes in rainfall patterns throughout the world.
- An increase in the frequency of tropical storms in many areas.
- An increase in the number and frequency of droughts.
- An increase in the process of desertification in arid and semi-arid areas.
- Increased flooding due to increased rainfall.
- Changes in climates throughout the world. Some places are experiencing milder winters and wetter summers.
- Wind patterns and ocean currents are thought to have changed in some areas.

Regional scale: equatorial and savanna regions of Africa

Some questions ask you to describe and explain the origin, nature and weather characteristics of Tropical Maritime (mT) and Tropical Continental (cT) air masses which affect West Africa.

When answering, mention **air masses**.

Air masses

- Air masses are a widespread expanse of air travelling horizontally throughout the earth. They are termed homogenous because the temperature and humidity of the air mass are similar throughout the surface layers.
- Horizontally they can measure up to hundreds of kilometres. Areas where the air masses are created are called **'source regions'**.
- Air masses which affect West Africa originate where uniform conditions are found, such as in tropical high pressure zones.
- If the air mass originates in a warm, dry area, for example over a desert, it will bring that kind of condition to the areas over which they pass.

- If it originates over water areas, the air mass will bring wetter conditions.
- The masses are generally termed **Maritime M** or **Continental C** depending on whether they originated over sea or land respectively.
- It is quite normal to assume that tropical air masses include any air which has its source in equatorial regions.

What are inter-tropical convergence zones and zones of 'Convergence and Divergence'?

- A zone of **convergence** is where winds meet and a zone of **divergence** is an area where the winds go in different directions.
- Converging winds include the trade winds which meet at the equator in a zone termed the **inter-tropical convergence zone (ITCZ)**.
- Two zones of **divergence** occur in the sub-tropical high pressure zones where the winds are usually fairly light. The zones are referred to as the 'horse latitudes'.
- In these zones, and in the ITCZ, the movement of air is vertical or convectional.
- In the ITCZ the air tends to rise along the inter-tropical front.
- The zone of convergence moves **northwards** during June/July and **southwards** during December/January with changes in the sun's angle of declination during the seasons.
- This shift in the ITCZ affects the climate of areas in these latitudes, especially rainfall patterns.
- When air masses from different source regions meet, the air at the edge begins a process in which the colder air forces warm air upward and **condensation** takes place in the upper parts.
- The place where this occurs is called the **front**.
- Where the trade wind belt of the northern latitudes meets the trade wind belt of the southern latitudes within the equatorial belt an **inter-tropical front** is formed.
- The weather associated with this front very much depends on whether the front has formed over the oceans or the continents.
- Air masses converging towards the inter-tropical front over oceans are **moist** in the **lower layers** and relatively **dry** at **higher levels**.
- At convergence there is some instability and large cumulus clouds appear and eventually this leads to intense shower and thunder conditions.
- Within equatorial areas the vertical movement of air through convection produces **convectional rain**.
- The climate of equatorial Africa is therefore best described as **high rainfall throughout the year** with **high temperatures with a small annual range** as shown in Figure 1.8.

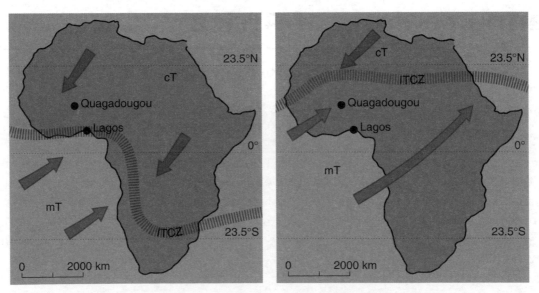

FIGURE 1.7: SELECTED AIR MASSES AND FRONTS OVER AFRICA – JANUARY AND JULY

What are the main features of an equatorial climate?

Figure 1.8 shows an equatorial climate.

When describing the main features of an equatorial climate, refer to:

1. **Annual temperatures** which are quite high.
2. The **annual temperature range** of between 3 and 8 degrees C is greater than that of the hot desert climate.
3. The main characteristics consist of **very little variation** in the high temperatures throughout the year with a maximum of around 32 degrees C and a minimum of 28 degrees C.
4. **Rainfall** throughout the year totals 2000 mm with a monthly maximum of 290 mm and a minimum of 175 mm.
5. There is **no seasonal variation in rainfall or temperature except** further north where conditions, especially rainfall, can vary.
6. Areas lying approximately 5 to 15 degrees north and south of the equatorial climate have a tropical or savanna climate.

What are the main features of a savanna climate?

Figure 1.9 shows a savanna climate.

When describing the climate, refer to **three distinct temperature seasons**, namely:

1. **A cool, dry season** which coincides with the low-sun period with temperatures ranging from 27 to 32 degrees Celcius.
2. **A hot, dry season** which immediately precedes the rains where temperatures can reach up to 38 degrees Celcius.
3. **A hot, wet season** which coincides with the high-sun period.

Note that the **annual range of temperature** in a savanna climate is **greater** than that of an equatorial climate.

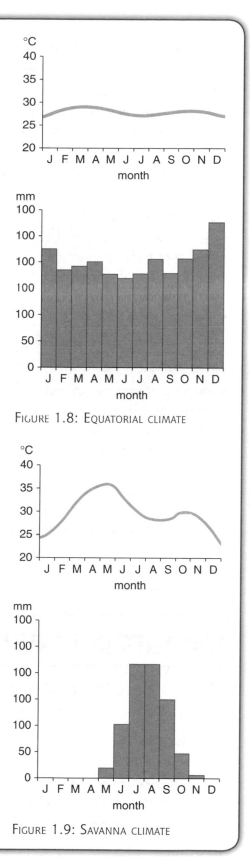

FIGURE 1.8: EQUATORIAL CLIMATE

FIGURE 1.9: SAVANNA CLIMATE

Quick Test

Read through the bullet points on inter-tropical convergence zones and zones of 'Convergence and Divergence', and write a sentence for each highlighted word.

How do air masses and inter-tropical fronts on the equatorial and savanna areas affect their climates?

Equatorial climate

- The climate is greatly affected by the convergence of the trade winds from the north and south.
- This convergence creates a region with a low pressure gradient with an upward movement of air that leads to convectional rain throughout the year.
- Temperatures remain high due to the latitudinal position of the sun giving maximum solar insolation throughout the year.

Savanna climate

- As the sun's vertical rays move north or south, depending on the time of year, the **doldrums and high-pressure belts** both affect the climate of savanna regions during different seasons.
- During the high-sun period the **doldrums and convectional effect** will predominate and this will be the **season of heaviest rainfall**.
- During the low-sun period, the drier air of the sub-tropical high-pressure belt will produce a **dry season**.

How and why are there variations in rainfall in West Africa?

These variations are affected by the following:

- In the northern hemisphere from March to July the ITCZ moves northwards across West Africa bringing heavy rainfall due to convectional rain.
- This rainy season lasts only about two months.
- At the **northern edge conditions are much drier** where the air meets dry air from the interior of the continent.
- As the ITCZ moves further southwards in winter, the drier continental tropical air is drawn southwards giving **drier** conditions in **the north western areas**.
- In some years there has been an apparent shift in the movement of the ITCZ which has resulted in **less rainfall than normal** in the areas north of the equator.
- This combined with a southward extension of the sub-tropical high pressure from the Sahara has resulted in long periods of drought.

Geographical methods and techniques

How to describe and explain climate graphs

Describing a graph involves commenting on temperature and rainfall patterns throughout the year.

Temperature

- Note the **highest and lowest temperatures** and the difference between them namely, the range.
- Note the **warmest and coolest periods** of the year.
- Note whether there are **definite seasons** as indicated by temperatures.
- Note whether the climate is **cold, cool, temperate, warm or hot**.

Rainfall

- Note the **distribution of rainfall** and when the **wettest and driest** seasons occur.
- Note the **amounts** of rainfall in each month, indicating the general pattern, e.g. **wet, dry, very dry**.
- Note the **overall amount of rainfall** throughout the year indicating the overall pattern, e.g. **very wet, moderate, dry, etc**.
- From your knowledge of different climate graphs you should be able to identify equatorial and savanna climates.

Sample question, answer and comments

Question

Explain the physical and human factors that might have led to the changes in global air temperatures shown in the diagram.

14 marks

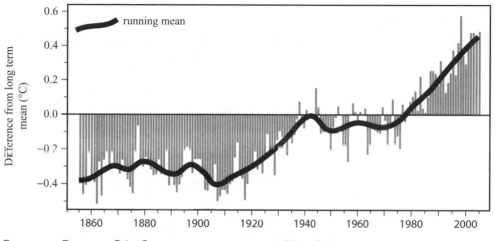

REFERENCE DIAGRAM Q1: GLOBAL AIR TEMPERATURES 1855–2005

Answer (1) denotes correct point

Physical factors which have led to global temperature change are the variation in solar energy, for example sun spots (1) and volcanic eruptions (1) as they produce dust which absorb solar energy therefore cooling the earth (1).

Human factors are deforestation (1) as the burning of trees cause CO_2 in the atmosphere (1). The burning of fossil fuels (1) releases CO_2 into the atmosphere. CFCs found in refrigerators (1) contribute to gases in the atmosphere. Methane (1) produced by cattle (1) contributes to gases in the atmosphere. Rotting vegetation also release harmful gases which contribute to an increase in the Greenhouse Effect (1).

Comments

This answer has sufficient points to achieve more than half marks, **10/14**. More marks could have been obtained if the answer had had more detail on the physical factors and had referred to industries and transport burning fossil fuels.

Topic glossary

Albedo: This refers to the amount of reflectability of surfaces on the earth such as land, ice and water.

Climate: The average weather conditions, usually taken over a period of 35 years.

Desert: Any area which has very low rainfall throughout the year. Note that there are cold and hot deserts in the world.

Drought: This occurs when there is a long period without rainfall. Notice that droughts do not necessarily just happen in desert areas.

Front: The boundary between two air masses. If the air on one mass is warmer than air being replaced the front is termed a **warm front**. If the air is colder than the air being replaced the front is termed a **cold front**.

Homogeneous air mass: A large volume of air which is composed of the same properties.

Insolation: This is the amount of heat taken in from the sun.

The Physical Environment

Inter-Tropical Convergence Zone (ITCZ): It was discovered that air converges not at a 'front' but in a broad zonal trough of low pressure in equatorial latitudes.

Latitude: This is the distance between the equator and the poles and is measured in degrees.

Pressure belts: Patterns of atmospheric circulation systems of either high or low atmospheric pressure.

Rainfall pattern: This refers to the distribution of rainfall throughout the year.

Range of temperature: This is the difference between the highest temperature and the lowest temperature in a year.

Solar energy: Any form of energy originating from the sun.

Hydrosphere

Hydrology is the study of the water within the earth, whether it is in the atmosphere, on the surface or underground. The movement of that water, the impact of it on the land and how this movement may be interrupted are also important aspects of this particular topic.

Hydrological cycle

You should be able to draw a diagram to show the global hydrological cycle, and name and describe the main elements within the diagram.

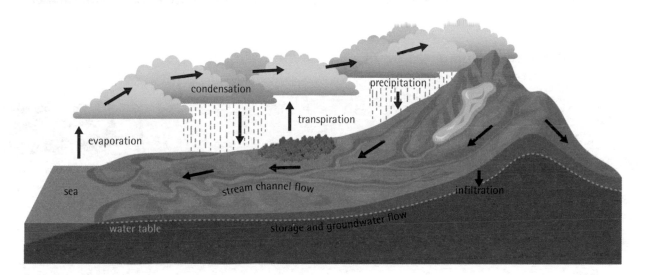

FIGURE 2.1: HYDROLOGICAL CYCLE

- Water exists on the surface in the form of oceans, seas, lakes, rivers and streams.
- It also exists in the atmosphere as rain and water vapour and underground as seepage within rock structures and in underground streams and lakes.
- The surface water can pass into the atmosphere through **evaporation**. It can be carried by winds and eventually returns to the surface as rain or snow.
- Water also exists on the surface as ice or snow, e.g. at the poles or in high altitudes.
- Water on the land may be returned as **run-off** to oceans and seas through rivers or streams.
- This intricate process of the movement of water back and forward between land, oceans and the atmosphere is called the **Hydrological Cycle**.
- The hydrological cycle works as a closed system in that there is a definite amount of water in land and water areas.
- This amount remains constant.
- The system is powered by energy from the sun.

There is a continuous movement of water between the different parts of the system through the processes of evaporation, transpiration and precipitation.

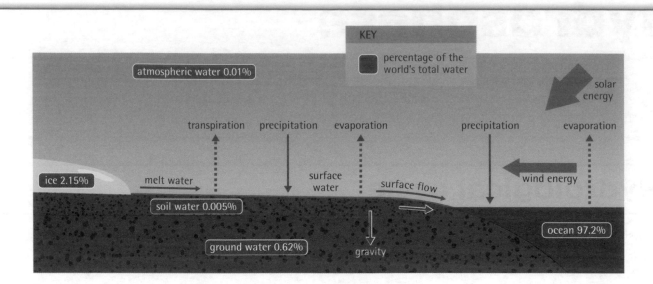

FIGURE 2.2: HOW BALANCE IS MAINTAINED WITHIN THE HYDROLOGICAL CYCLE

Balance is maintained when INPUTS = OUTPUTS.

Inputs:

- Rainfall
- Rivers and streams flowing into the sea
- Groundwater seepage into the sea

Outputs:

- Evaporation from the sea and rivers
- Transpiration from plants
- Evaporation from land surface

Top Tip

When discussing balance, remember the following equation: INPUTS = OUTPUTS

River drainage basins

Drainage basins are the total areas, known as catchment areas, which drain into rivers.

River drainage basins and water storage

The main characteristics of a drainage basin system include:

- **Soil water and ground water storage** which varies due to changes in **precipitation, evaporation, transpiration, infiltration and local geology**.
- **Stored water** which is water held within the system as ice and snow, in lakes or in the soil.

Top Tip

Write a list of the terms highlighted in the section on drainage basins. Use these words in answers to questions on this topic.

Quick Test

Draw a labelled sketch of the hydrological cycle from memory.

Hydrographs

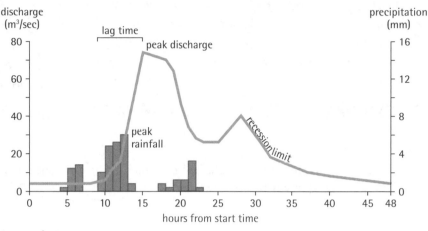

FIGURE 2.3: HYDROGRAPH

The rate of flow or discharge of water within a river basin can be measured or recorded by a graphic technique known as a **hydrograph** as shown in Figure 2.3. You should know the key features of a hydrograph.

You can be asked to describe and explain patterns shown on a river hydrograph. When doing this refer to:

- **Total rainfall:** This is the amount of rainfall which has fallen over a specific length of time – usually several days.
- **Time:** This is the time over which the run-off or discharge is measured and recorded.
- **Discharge:** This is the amount of water which is been discharged by the basin within the specified time scale and is measured in cumecs (m3/sec).
- **Rising limb:** This indicates how quickly waters begin to rise.
- **Peak flow:** This is the maximum discharge during a storm period.
- **Lag Time:** This is the difference in time between the time of peak/maximum rainfall and the peak discharge/highest level of the river.
- **Recessional or falling limb:** This indicates the speed at which the water level in the river declines after its peak flow.
- **Base flow:** This shows the normal level of a river.
- **Quick flow:** This is the water which is fed into the river due to overland run-off.

A **storm hydrograph** is one which displays **two** basic features:

Feature 1: The rainfall from a rain storm which is shown by a bar graph, and

Feature 2: The river discharge before, during and after the rain storm which is shown by line graphs.

In effect, the **storm hydrograph** indicates how a river responds to a rain storm.

How to analyse hydrographs

When analysing hydrographs, mention the following:

- During a storm, **discharge** does not increase immediately since only a small amount of the rain will fall directly into the river channel.
- Water will reach the river from the overland flow.
- This is the **surface run-off** and it will subsequently be supplemented by water from **throughflow**.
- These are both shown on the graph. Note these in your analysis.
- The **rising limb** shows the overland flow and the **falling limb**, which is less steep, indicates that there is still water in the system from throughflow, surface water and water in stream channels.

The Physical Environment

- Note these limbs in your analysis.
- Note the **lag time and discharge**.
- Rivers which have a **short** lag time and a **high** discharge are more likely to **flood** than rivers with a **long** lag time and a **low** discharge.
- Depending on the characteristics of river basins in terms of size, shape, relief, geology, etc., hydrographs for two basins receiving the same amount of rainfall can be very different.

How to compare hydrographs for two different river basins

When comparing, note the following:

- Drainage basin **A** has a much higher density than that of drainage basin **B**. Comparison of the hydrographs for each basin shows that for basin A the lag time is much shorter (30 hours) than basin B (55 hours).
- The peak discharge for A is much higher (140 cumecs) than basin B (50 cumecs).
- The rising limb for B is much less steep than for A as is the recession (falling) limb.
- Apart from the difference in density of drainage within the two basins, it is quite possible that there is some considerable variation in the relief pattern where the pattern may be much steeper in basin A than B.
- The soil and rock types may also vary, thereby affecting the overland flow patterns and perhaps the throughflow patterns.
- There may be some differences between the vegetation present within the two basins. For example, there may be more woodland in B than A which will affect discharge.

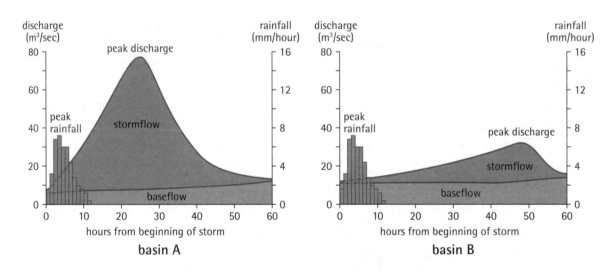

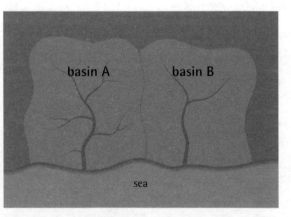

FIGURE 2.4A: TWO SELECTED FLOOD HYDROGRAPHS/RIVER BASINS, B: URBAN/RURAL HYDROGRAPH

- The net result of these differences in hydrographs and basin characteristics is that the river in basin A is more likely to flood than the river within basin B.
- Without a lot more detail on the characteristics of both systems flood predictions cannot be wholly accurate.

In the 2008 examination, candidates were asked to compare urban and rural hydrographs by describing and explaining differences. If asked this in a future exam you could refer to the following:

Differences could include:

Interception

In rural areas there is a longer lag time because vegetation will intercept precipitation and therefore prevent the water reaching the soil and river quickly. In urban areas the lag time is shorter due to concrete/tarmac and drains which carry the water much more quickly to the river system.

Surface run-off

The rising limb is much steeper in the urban hydrograph because, unlike rural areas where water overflows into fields in the flood plain, the water courses are lined and embanked which speeds up the flow of water.

Storage

The falling limb on the urban hydrograph is much steeper because of a lack of infiltration and underground storage of water. Water in rural areas continues to flow into the river and underground and therefore the base flow is much slower with a more gentle falling limb.

What are the main factors which affect flooding?

Physical factors include:

- Extreme weather conditions such as torrential rain, continuous rain for several days and heavy snowfalls which melt while it is raining.
- The amount of water being discharged by a river can increase due to these conditions causing the river to overflow its banks, leading to widespread flooding.
- High summer temperatures can make the ground hard which reduces infiltration when it rains.
- Ground can become impenetrable due to freezing conditions in winter.
- Surface run-off over areas with permeable or impermeable rock.
- Whether the land is steeply or gently sloping which can increase or decrease the rate of run-off.
- Lag time is also a factor. Rivers with a short lag time and a high discharge are more likely to flood than those with a long lag time and a low discharge.

Human factors include:

- More surface areas being covered by tarmac and concrete thus reducing infiltration in cities and towns.
- Not enough drains being laid to reduce flood waters.
- Drains becoming clogged by litter and falling vegetation.
- Arable farming can expose more soil.
- Removal of natural vegetation such as forests and woodland. Less rainfall is therefore intercepted and run-off is increased.

Quick Test

Without looking at the last section, list the main items you should refer to when answering a question on a given hydrograph.

Fluvial landforms and landscapes

River processes

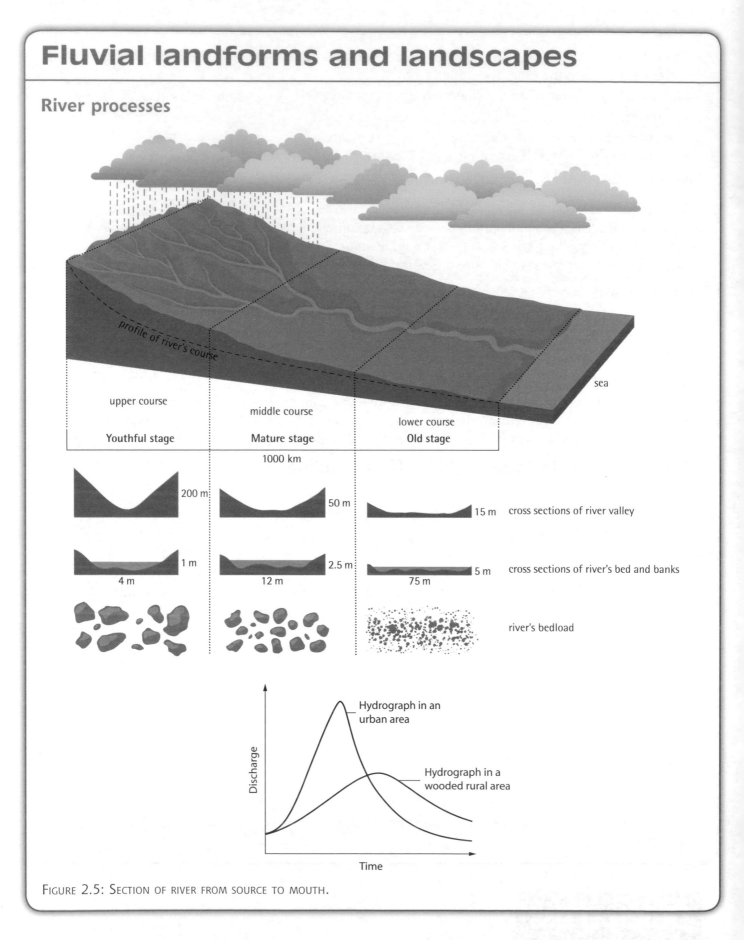

FIGURE 2.5: SECTION OF RIVER FROM SOURCE TO MOUTH.

What are the effects of flowing water in relation to the processes of erosion, transportation and deposition in each of the three stages of a river valley?

These three stages include:

Stage 1 (Youthful stage), Stage 2 (Mature stage) and Stage 3 (Old Age stage)

What are the main features of each stage?

Stage 1

- The river is shallow, narrow, with rapids, waterfalls, interlocking spurs, boulders and large stones.
- The valley is steep sided – V shaped.
- The main process is downward erosion.

Stage 2

- The river is wider than stage 1.
- The valley has a flood plain, meanders and more gently sloping sides.
- The gradient is less steep than stage 1.
- There is almost a balance between erosion and deposition.
- Deposits consist of smaller stones and gravel.

Stage 3

- The river is much deeper and may be tidal.
- The valley is very broad and flat with gentle slopes.
- The features will include meanders, ox bow lakes, levees and river terraces.
- The main process is deposition with sand and silt being deposited.

Effects of flowing water include:

- As water flows downwards under the force of gravity it follows the path of least resistance. As rivers flow, three separate processes take place, namely **erosion, transportation and deposition**.

Erosion

- **Erosion** is the wearing away of the land surface over which the river flows. This wearing away occurs in different ways. It occurs through processes such as **abrasion, corrasion, hydraulic action, attrition and corrosion**.
- **Abrasion** happens when small particles of rock carried by the river rub the suface of the river bed and wears away this surface.
- **Corrasion** happens when there is abrasion of the river bed by material carried along by the river.
- Water flowing past river banks can be forced into cracks and after some time the banks can collapse. This process, which is more gradual than abrasion, is called **hydraulic action**.
- **Attrition** occurs when water flow causes pieces of rock to collide with each other and break down.
- **Corrosion** is caused by chemical reactions between carbonic acid or acid from vegetation and the river bed.

Transportation

- **Transportation** is the second process which is carried out by rivers.
- The material which is eroded by the processes outlined above is carried or transported by the river and this material is called the **river's load**.

Deposition

- **Deposition** occurs when the velocity of the river begins to decrease. When this happens the river no longer has the competence to carry all of its load.
- If a river overflows its banks during a flood, material is deposited creating a feature called the **flood plain**.

River landforms

- If a river flows over hard rock and then over a band of less resistant rock, the less resistant rock will be worn away much more quickly than the overlying rock.
- Eventually so much of the underlying rock may be eroded that there will be nothing left to support the rock above.
- The overlying rock then collapses, forming a **waterfall**.
- As the process is repeated the waterfall will retreat and this may eventually lead to the formation of a **gorge**.
- A gorge is a deep valley with very steep sides and a narrow valley floor.
- In **the middle** stage of the valley, the valley sides are less steep although they may still be hilly. The gradient becomes more gentle and the width of the valley increases with an increase in flat land along the sides of the river.
- River bends or meanders begin to appear as the river finds the course of least resistance.
- The speed of flow of the river varies across the meander.
- The rate of flow is much **faster on the outside bend** of the meander and at this point the water is eroding the outer bank.
- Speed is **slower on the inside bend** and the river begins to deposit its load.
- In the **lower or final stage** of the valley the river widens and flows more slowly across the land.
- The **river flood plain** usually increases in width and large meanders are common.
- Material may be deposited in the middle, forming islands.
- This process is known as **braiding**. The islands of deposited material divided by the different channels are called **'eyots'**.
- As the sizes of meanders increase, eventually the river may cut a channel between the narrowest point of the bends and the feature which is left cut off from the river is called an **oxbow lake**.
- Figure 2.7 shows stages in the formation of an oxbow lake.

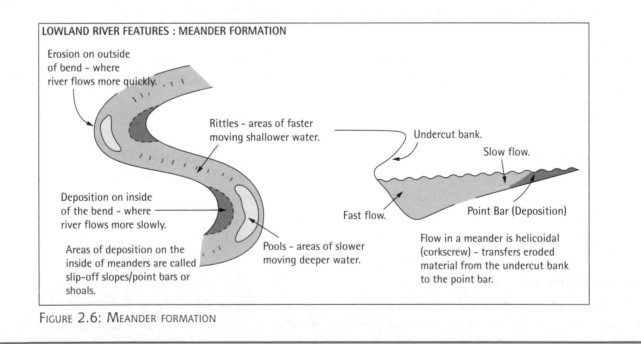

LOWLAND RIVER FEATURES : MEANDER FORMATION

Erosion on outside of bend - where river flows more quickly.

Rittles - areas of faster moving shallower water.

Undercut bank.

Slow flow.

Deposition on inside of the bend - where river flows more slowly.

Fast flow.

Point Bar (Deposition)

Pools - areas of slower moving deeper water.

Areas of deposition on the inside of meanders are called slip-off slopes/point bars or shoals.

Flow in a meander is helicoidal (corkscrew) - transfers eroded material from the undercut bank to the point bar.

FIGURE 2.6: MEANDER FORMATION

Quick Test

Describe the main features of the upper stage (youthful) and the lower stage (old-age) of a river valley.

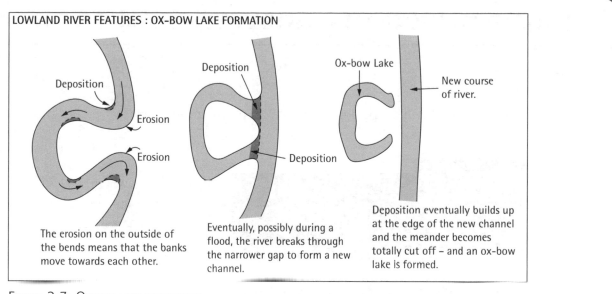

LOWLAND RIVER FEATURES : OX-BOW LAKE FORMATION

Deposition

Erosion

Erosion

The erosion on the outside of the bends means that the banks move towards each other.

Deposition

Deposition

Eventually, possibly during a flood, the river breaks through the narrower gap to form a new channel.

Ox-bow Lake

New course of river.

Deposition eventually builds up at the edge of the new channel and the meander becomes totally cut off – and an ox-bow lake is formed.

FIGURE 2.7: OXBOW LAKE FORMATION

- **Deltas** are formed by the process of deposition at the mouth of the river as the river loses energy. This can only happen if there is no strong ocean current.
- **Estuaries** are formed where rivers broaden out gradually as they meet the sea.

Geographical methods and techniques

Rivers on Ordnance Survey maps

Using an OS map, you should be able to describe the main characteristics of a river and its valley.

The basic system of surface drainage is known as a **river basin**.

When describing the features of a river and its valley refer to:

- The stage of development of the valley, e.g. young, mature or old as determined from the contour patterns of the valley.
- Direction of flow should be noted as indicated by both contour lines and spot heights at various points along the course.
- The gradient should be described in terms of whether it is steep or gentle using spot heights and contours when deciding on steepness.
- Note distinct features within the valley, e.g. meanders, tributaries, oxbow lakes, flood plain width.
- Refer only to physical features and not human features such as bridges, roads or settlement.
- If referring to certain physical features give appropriate 6 figure grid references.

Top Tip

Use the following structure as a basis for your answer: Refer to: Stage: Direction: River gradient: River features: Valley floor width: Slope of valley sides: Valley shape. Give appropriate grid references.

Quick Test

Draw a series of labelled diagrams to show how an oxbow lake is formed.

Sample questions, answers and comments

Question

(a) Study Reference Diagram Q1A.

Describe and explain the hydrographs for the River Severn and the River Wye after the storm of 6 August 1973. **8 marks**

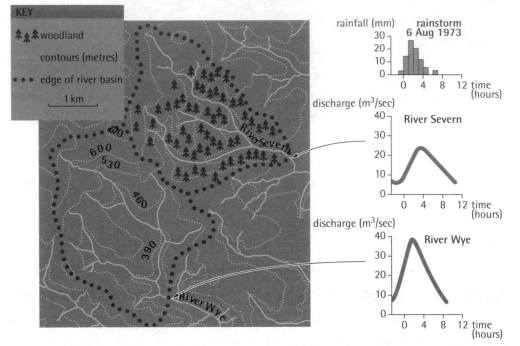

REFERENCE DIAGRAM Q1A: HYDROGRAPHS OF THE RIVERS SEVERN AND WYE

(b) Study Reference Diagram Q1B.

Select one of the flood plain features shown and, with the aid of a diagram or diagrams, describe and explain its nature and formation. **8 marks**

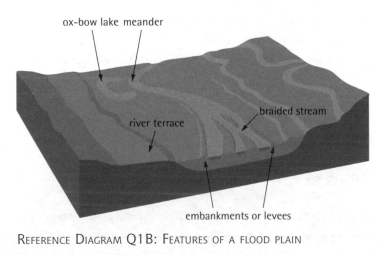

REFERENCE DIAGRAM Q1B: FEATURES OF A FLOOD PLAIN

Quick Test

Write a sentence for each of the words in the Top Tip to show that you understand what they refer to.

Answer (1) denotes correct point

(a) *The River Severn basin is covered by woodland (1). This would delay the water from reaching the river basin as quickly as the water in the Wye basin (1) as this is not covered by vegetation. The woodland and vegetation would stop the water from running down the valley sides quickly and also some would be taken up by the roots (1). This accounts for the longer delay time at the beginning of the River Severn's hydrograph (1). The vegetation is also responsible for the lower height reached by the Severn – 23 m3/s compared to the high 47 m3/s in the Wye basin (1). The River Severn's drainage basin is also situated on gentler slopes than the Wye. This factor also reduces the run-off in the Severn's basin (1). The River Wye takes a shorter period of time, however, to return to its normal discharge level – about 9 hours (1). This is due to the water being held back longer in the Severn's basin.*

Comments

The first part of the answer correctly compares both basins in terms of the rate of discharge and relates this to the presence of woodland in the River Severn basin, thus gaining **2 marks**. The reference to lack of vegetation in the Wye basin is not correct. The answer then tends to repeat the explanation of the impact of the woodland although the point about the tree roots taking up water would receive an **additional** mark. A **further** mark is obtained by relating this last point to the delay time in the River Severn's hydrograph.

The answer makes references to the discharge rates, using data from the graphs although the reference to 47 m3/sec for the Wye basin is incorrect. This obtains a **further** mark. There are no marks for the statement on the slopes of the Severn basin since the contour pattern on the map does not confirm this.

1 final mark is obtained for the descriptive point relating to the shorter time for the Wye to return to its normal discharge, again using correct calculations from the graphs. In total the answer would probably merit **7 out of a possible 8**. A very good answer.

(b) *In a meander the river flows fastest on the outside curves (1). This means that more erosion occurs on the banks of the outside curves than on the inside ones (1). This means that the neck of the meander will slowly get closer and closer together (1). See diagrams 1 and 2. Eventually the neck gets so close that during a storm or event that causes the river level to rise the river will 'punch' (1) through leaving the curve out (1).*

Comments

The first two comments on the speed of the river on the outside curves and the fact that erosion takes place on the outer bend are both worth 1 mark each. The reference to the neck becoming closer a shown in the diagram is worth a further 1 mark.

The final marks are obtained from the remarks on the neck closing during a rise in river level and finally 'punching' through and isolating the curve. The description of the formation process could have been a little more precise, especially in reference to the creation of the ox bow lake in the final stage. Altogether the answer would probably merit 5 x 1 marks, thus 5 marks out of possible out of 8. Just above a basic pass.'

Topic glossary

Evapotranspiration: The process by which moisture is returned to the atmosphere by direct means through evaporation combined with transpiration from vegetation.

Flood plains: These are areas formed when rivers overflow their banks and flood surrounding flat land, leaving deposits of silt until eventually a flat flood plain is built up.

Hydrosphere: The name given to all of the water surfaces on the earth.

Infiltration: The process by which water seeps into the soil and sub-soil.

Interlocking spurs: The bottom part of slopes in a valley which intertwine.

The Physical Environment

Lag time: A time delay between the arrival of a signal in a meteorological measuring instrument and the response of that instrument.

Meanders: These are large bends which form as the river nears its mouth.

Oxbow lakes: These are formed when the bend of a meander is cut off forming a crescent-shaped lake.

Precipitation: All forms of moisture in the atmosphere including rain, hail, sleet and snow.

Quickflow: The surface movement of water, from precipitation, which is not interrupted by vegetation and which runs as a shallow, unchannelled sheet across the soil.

Tributary: A small stream/river which runs into a larger river.

Lithosphere

The lithosphere is the name given to the variety of **physical landscapes**, specifically those relating to (for the purposes of the Higher Geography course and exam) the United Kingdom and the processes which create them. The types of landscape which you are required to know about include glaciated uplands, areas of upland limestone and coasts. What are the main features of these landscapes and how were they formed? We can begin with glaciated uplands.

Glaciated uplands

Landforms include:

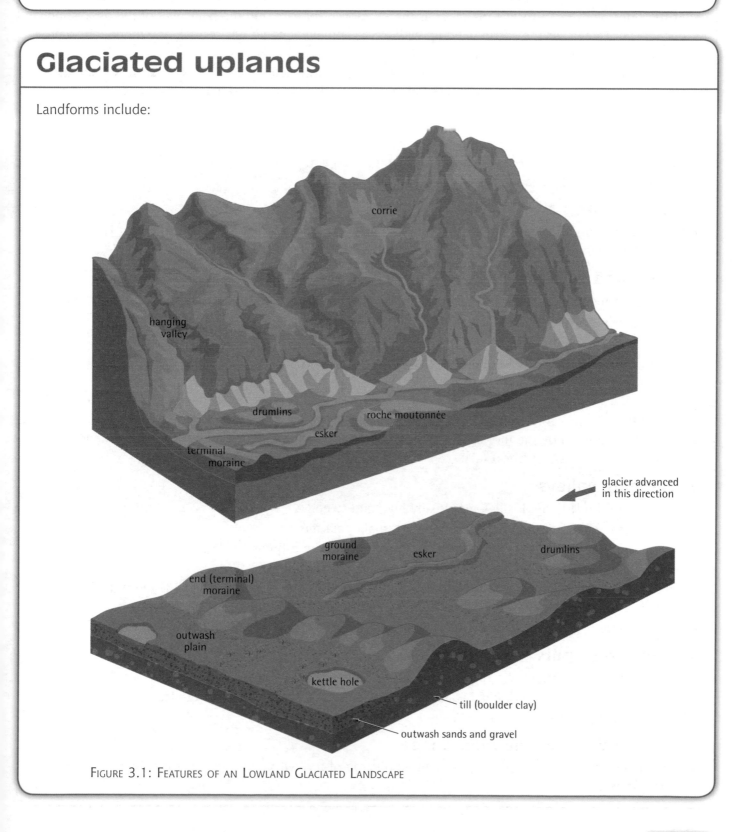

FIGURE 3.1: FEATURES OF AN LOWLAND GLACIATED LANDSCAPE

The Physical Environment

1. Corries, Arêtes, Pyramidal peaks

Corries

- Corries are steep-sided hollows in the sides of mountains where snow accumulated and gradually compacted into ice. The rotational movement of ice in the hollow caused considerable erosion both on the floor and on the sides of the depression.
- The erosion on the floor was caused by **abrasion** and the floor became **concave** in shape; the edge took on a ridge-shaped appearance.
- At the sides plucking of rocks took place as the ice moved forward and the back wall of the depression became very steep. As the corrie filled up with ice, eventually it could not contain any more and some of it moved down the slope to a lower level.
- The weight of the ice caused the glacier to slide on top of the melting ice. This is called **basal sliding**.
- Occasionally as the ice melted, meltwater filled the corrie forming a corrie lake called corrie lochs (Scotland) and tarns (England).

Arêtes

- Often corries developed on adjacent sides of a mountain and when they were fully formed they were separated by a 'knife'-shaped ridge called an arête.

Pyramidal peaks

- If corries develop on all sides of a mountain, the arêtes will form a jagged peak at the top. This feature is called a pyramidal peak. These are further sharpened by **frost action**.

2. U-shaped valleys

- As a glacier moved downhill through a valley, the shape of the valley was transformed. A material called **boulder clay** was deposited on the floor of the valley.
- As the ice melted and retreated the valley was left with very steep sides and a wide flat floor.
- A river or stream may flow through the valley due to meltwater from the glacier.
- This replaced the original stream or river and is termed a **'misfit'** stream.
- The material which was pushed in front of the glacier, and left as the glacier melted, is called **terminal moraine**.
- The valley dammed by the moraine may then flood creating a lake which may twist and turn and therefore it is termed a **ribbon lake**.

3. Hanging valleys

- The sides of the U-shaped valley are usually high and steep.
- During the ice age, tributary valleys often had smaller glaciers.
- The glacier in the main valley cut off the bottom slope of the tributary valley leaving it high above the main valley.
- Tributaries of the main valley will therefore plunge from the slopes of the main valley into the bottom of the valley.
- These smaller valleys are called **hanging valleys**.

4. Truncated spurs

- A spur is the bottom part of a slope which juts out into the main valley.
- As the ice cut through the original valley, the original spurs were removed by the ice.
- The feature remaining once the ice melted is called a truncated spur.

5. Crag and tail, roche moutonnée

- These features occur where the glacier met an outcrop of rock which was harder than the rock of the surrounding area.

- The sides of the harder rock (crag) are worn away by the glacier but the land behind the crag is protected and forms a long, gently sloping ridge called the tail.
- **Roches moutonnées** are outcrops of harder rocks which have been smoothed on the side facing the ice to give a gentle slope and plucked on the other side to produce a more jagged, steeper slope.

Features of glacial deposition

These include:

Moraines

- Moraine consists of material known as 'boulder clay and till' which has been eroded and transported and deposited by the glacier. This material may be dumped at the end or snout of the glacier and is called **'terminal moraine'**.
- Those formed by material dumped at the sides or in the middle where two glaciers came together are called **'lateral'** and **'medial'** moraines respectively.

Erratics

- These are large boulders which have been lifted, carried and deposited by the glaciers some distance away in a different part of the country. The rock type of the erratic is usually different from the rocks which are common to the area in which has been deposited.

Outwash plains

- These are gently sloping plains consisting of sands and gravel. These have been deposited by meltwater streams flowing out from the ice sheet and carrying material collected by the glacier.

Eskers, Drumlins, Kames

- Eskers are elongated ridges of coarse, stratified, fluvioglacial sands and gravels and are thought to have been formed by meltwater tunnels within the lower parts of the glacier which deposited the material.
- Drumlins are oval-shaped mounds which can be up to 100 metres high and have a 'basket of eggs' look to them.
- The material in them was deposited due to friction between the ice and the underlying rock causing the glacier to drop its load.
- Kames are irregular-shaped mounds of material consisting of sands and gravel again laid by glacial streams.
- Sometimes they form terraces on the side of the valley, where the streams ran along the sides of the ice trapped against it by the valley walls.

Formation processes

During the last 2·5 million years the British Isles have been covered at different times with large sheets of ice.

- Glacial erosion occurs through a process called **abrasion** (a sandpapering effect as the ice moves across the land) and **plucking**, where pieces of rock are torn away from the land.
- The weight of the ice causes the glacier to slide on top of the melting ice. This is called **basal sliding**.
- The rate of flow of the glacier depends on the type of rock over which it flows, the amount of ice in the glacier and the slope of the land. This process of melting is called **ablation**.
- Meltwater flowing from the glaciers further erodes and deposits material in a process called **fluvioglacial processes**.

Top Tip

Study the diagram showing contour patterns of various upland glaciation features and try to find similar features on an OS map of a glaciated area.

In the exam you may be given an OS map or a diagram and be asked to identify features of glacial erosion and to explain how selected features were formed.

If asked how the landscape was formed you can refer to the processes of glacial erosion.

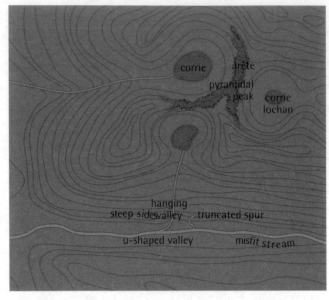

FIGURE 3.2 CONTOUR PATTERNS: GLACIATED UPLANDS.

Top Tip

Study the sketch in Figure 3.2 and learn to recognise the various features shown.

Upland limestone

You may be asked in a question to identify from a diagram or map surface and underground features of upland limestone and explain how they were formed. Figure 3.3 shows a summary diagram of the features associated with areas of upland limestone.

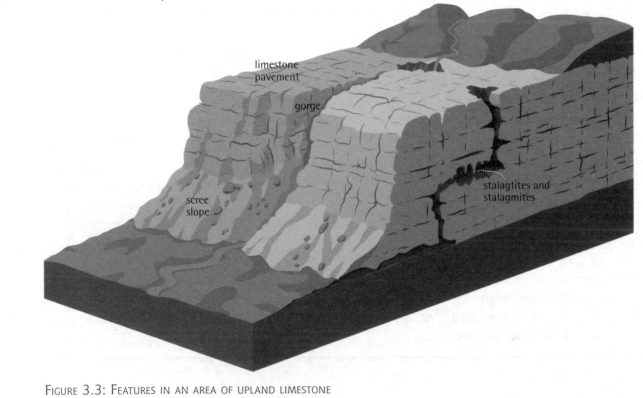

FIGURE 3.3: FEATURES IN AN AREA OF UPLAND LIMESTONE

Limestone is a sedimentary rock consisting mainly of calcium carbonate – at least 80%. Carboniferous limestone was formed about 250 million years ago and the landscapes have specific features which are recognisable including:

- Surface features such as limestone pavements (clints and grykes), intermittant drainage, pot/swallow holes, scarp slopes and gorges.
- Underground features such as caverns, stalagmites and stalactites and underground lakes.

Surface features

What are limestone pavements and how were they formed?

- Through glaciers passing over the top of an upland limestone area, the top-soil was removed leaving an area of exposed rock.
- Through the subsequent chemical action of rainwater dissolving the limestone, joints widened and deepened on the surface creating **large blocks resembling pavements**.
- The cracks or fissures between the blocks are called **grykes** and the large blocks separated by the grykes are called **clints**.
- Pot holes/swallow holes are formed where persistent widening of a major joint occurred, possibly by a stream disappearing underground.

What is intermittent drainage?

- Intermittent drainage occurs in limestone areas when streams, which drain areas of impermeable rock, carry on into the limestone area and disappear through the permeable limestone. The drainage is thus interrupted.

How are gorges formed?

- Gorges are formed when the underground cavern can no longer support the rock above. The rock collapses leaving a gap called a gorge.
- They are also formed through the erosional effects of glacial meltwater.

Scar slopes

- The edge of the limestone area which is exposed to the elements can form a steep slope known as a **scar slope** through the process of frost shattering.
- Water seeps into little cracks on the slope surface and freezes during cold winters splitting the rock.
- Fragments of rock fall down the slope and accumulate at the bottom.
- This is known as scree.
- Due to the lack of water, vegetation is sparse or non existent in limestone areas.
- Exposed hard grey limestone is clearly seen on the surface.
- This landscape is sometimes referred to as **Karst scenery**.

Undergound features

Underground caverns

- As the process of dissolving the limestone continues underground sections of the rock may collapse onto the bedrock creating **underground caves**.
- As the surface water meets the impermeable underground rock, this leads to the formation of **underground lakes and streams**.

What are stalagmites and stalactites?

- Stalactites and stalagmites are also formed underground in caverns.
- Dissolved limestone deposits have water which evaporates due to air currents in the cave.
- These deposits of limestone are left either hanging down from the ceiling of the cave called stalactites or building up from the ground called stalagmites.

Coastal landforms

In some examinations you may be asked to identify certain coastal landforms and explain how they were formed. These include:

cliffs, wave cut platforms, caves, arches, stacks, headlands, bays, spits, bars and, longshore drift.

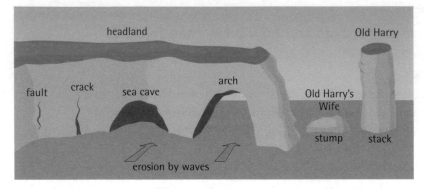

FIGURE 3.4: A TYPICAL LANDSCAPE OF COASTAL EROSION CONTAINING HEADLANDS, CAVES, ARCHES AND STACKS

How were these features formed? The following processes were involved:

- **Corrasion** (abrasion) is caused by waves throwing beach material against cliffs.
- **Attrition** happens when waves cause rocks and boulders to break up into small particles by bumping them together.
- **Hydraulic action** is the process of waves compressing air in cracks in cliffs. As a result any weaknesses in rockface of cliffs and headlands are widened creating caves.

Cliffs

- Cliffs are formed by wave action undercutting land which meets the sea. This occurs at about high tide level.
- A notch is cut and as the land recedes the cliff base is deepened by wave erosion.
- At the same time the cliff face is continually attacked by weathering processes and mass wasting such as slumping which causes the cliff face to become less steep.
- When high, steep waves break at the bottom of a cliff, the cliff is undercut forming a feature called a **'wave cut notch'**.
- Continual undercutting causes the cliff to collapse eventually.
- When resistant rocks alternate with less resistant rocks along a coast and are under wave attack, the resistant rocks form **headlands** while the less resistant rock is worn away to form **bays**.
- Although the headlands gradually become more vulnerable to erosion, they protect the adjacent bays from the effects of destructive waves.

Caves, arches and stacks

- Caves are formed when waves attack cliffs with resistant rock along lines of weakness such as faults and joints.
- Corrasion, corrosion and hydraulic action by the waves widen the joints.
- Eventually the cave is widened and deepened by the waves.
- The waves undercut part of the cliff and can cut right through the cave to form an **arch**.
- Continual erosion causes the arch to widen and eventually the roof of the arch becomes too heavy to be supported and collapses to leave a piece of rock left standing called a **stack**.

Spits and bars

- **Spits** result from marine deposition and consist of a long narrow accumulation of sand or shingle with one part still attached to the land.
- The other end projects at a narrow angle into the sea or across an estuary. This end is often hooked or curved.
- **Bars** are ridges of sediment formed parallel to the coast and can be exposed at high or low tides.
- **Tombolos:** Often bars form barriers across bays. If a bar joins an island to the mainland it is called a tombolo.

Other coastal processes include:

Longshore drift

- This is the process by which waves move material up and down a beach.
- This material is usually deposited in a zig-zag fashion due to the effects of winds on waves.

Sea level changes

- This happened in post glacial periods when large amounts of ice melted and caused sea levels to rise, often drowning parts of coastlines.

Slumping

- This is the movement of surface rocks or superficial material which has become detached from a hillside or cliff face.

Rockfalls

- These occur when small blocks of rock become detached from a cliff face due to the sea undercutting the cliff along joint patterns.

Cliff line retreat

- This happens when the cliff face is gradually worn back by slumping, undercutting and rockfalls.

Identifying features of physical landscapes

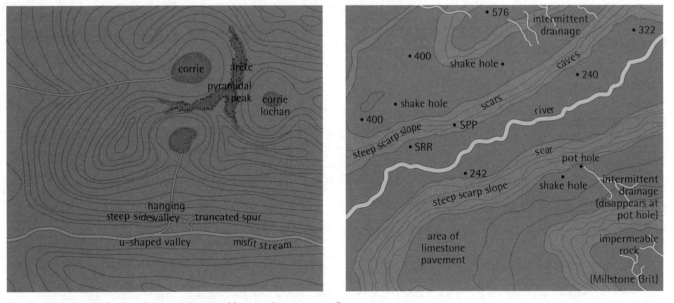

FIGURE 3.5 A; B; C: GLACIATED UPLANDS; UPLAND LIMESTONE; COASTS

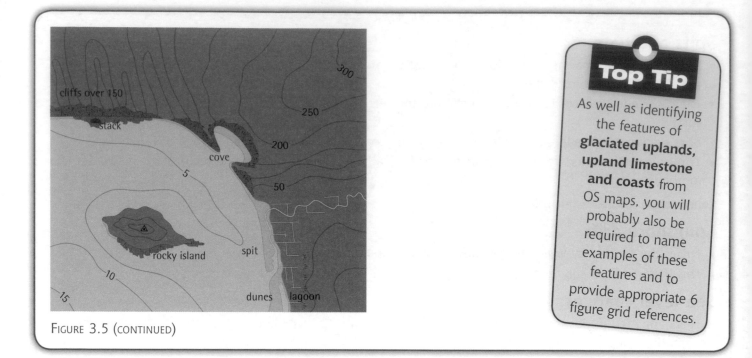

FIGURE 3.5 (CONTINUED)

Processes of slope formation

You should be able to describe and explain the processes of weathering and erosion which have led to the formation of physical features in the landscape.

Weathering involves the natural breaking down of rocks into smaller particles. There are basically two types of weathering, namely, **physical and chemical weathering**.

Physical weathering

This involves the disintegration of rocks into smaller particles without altering their chemical composition, for example through **frost shattering**, where

- Water which seeps into small cracks or fissures in rocks can freeze during sub-zero temperatures.
- As the temperature rises between 0 and 4 degrees (C) the ice expands and exerts pressure on the rock.
- Gradually pieces of rock break off and may fall down slopes under gravity, collecting at the bottom as scree.

Chemical weathering

Rainwater or water which seeps through vegetation such as moorland can become acidic. When this water comes into contact with rocks this can result in a chemical reaction in which the rock begins to break down.

Mass movements

Some questions ask you to describe and explain the conditions and processes which have led to the formation of rockfall and scree slopes and landslip especially as they affect coastlines. These processes are referred to as **mass movements**.

- In the new syllabus you will not be expected to answer a question on the general topic of 'mass movement'.

But

- It is important to have some knowledge of them since they affect other landscapes including areas of upland glaciation, upland limestone and coasts.

Mass movement of soil and rocks can range from minimal, almost imperceptible, movement such as soil creep to dramatic movements such as landslides. Other movements include scree formation and rock falls.

Factors influencing movements include:

1. The degree of slope

- Slopes which are gentle, no more than 5 degrees, are more likely to experience **soil creep**, whereby the soil gradually and slowly moves down the slope.
- Steeper, almost vertical, slopes will often have rock falls due to weathering.

2. Water content

- If the water content of the soil increases, for example during periods of heavy rainfall, the soil becomes saturated and may begin to move or slip over the bottom layers.
- This can lead to mud and earth flows.

3. Nature of underlying rock

- There are occasions where strong rocks are underlain by relatively weaker rock strata.
- If the underlying strata is worn away it will be unable to support the rocks above and this can result in landslides or **slumping**.

Sample question, answer and comments

Question

Study the OS map extract number 1056/OLM2: Ingleton.

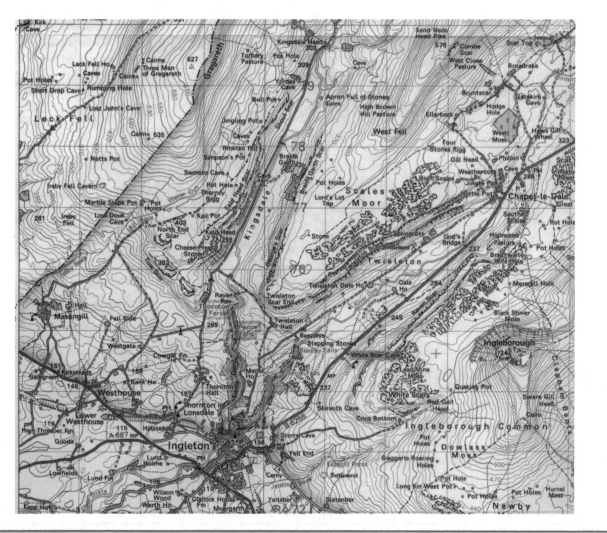

(a) The area of the Yorkshire Dales National Park, shown on the map, is characterised by Upland Limestone scenery.

Describe the evidence to support the above statement, referring to specific named features shown on the map extract.   24 marks

(b) Choose any one Upland Limestone feature (other than waterfalls), described in your answer to part (a), and, with the aid of a diagram, explain its formation.

12 marks

Answer (1) denotes a correct point

(a) *Limestone scenery can be found in Thistleton Scars (1) 725764. Shake holes (1) and pot holes where there are fissures in the limestone can be found in Rantry Hole 722732 (1) and Gritstone Pot 72673 (1). Disappearing stream can be found at 744723 (1) where water comes off impervious cap rock and disappears down fissures in the limestone (1). Limestone pavement (1) can be found at Green Edge 729744 (1).*

Comments

The first paragraph has references to four correct features with appropriate grid references and would be worth **8 marks**. The references to the 'disappearing streams',
limestone pavements and the location at Green Edge would merit a further **6 marks**.
The correct grid references would gain a further **2 marks** giving a total of **16 marks out of a possible 24 marks**.

Reference to other surface limestone features would have gained additional marks, but these were not given in the answer. An above average answer.

Answer

(b) *Limestone pavement is formed when glacier ice strips (1) away the overlying top soil leaving the underlying limestone bare (1).*

Rain water, being weak carbonic acid (1), begins to weather (1) away the joints of the limestone forming grooves called grykes.

The blocks between the grooves are called clints (1).

Comments

The candidate correctly indicates the work of the glacier in the formation process and 'leaving the underlying limestone bare', are sufficient for **2 marks**. The further statements on the chemical erosion of the limestone through 'weak carbonic acid in rainwater' and the correct identification of the joints and blocks – grykes and clints – would merit a further **4 marks** with a final mark for the diagram.

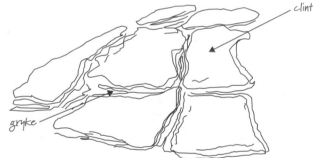

This would result in a total of **6 marks out of the possible 12 marks available**. A basic pass.

Topic glossary

Abrasion: This is the process by which rocks within ice sheets and rivers scrape and erode the land over which they pass.
Alluvium: This is the material deposited by a river, usually over its flood plain.
Arête: This is a narrow ridge between two corries and is formed as the corries are formed on two adjacent sides of a mountain.

Attrition: This process occurs as rocks in a river wear away by constantly rubbing together.

Corrie: This is an armchair-shaped hollow on the side of a mountain which was formed by ice filling a hollow and eroding the side of the mountain by abrasion and plucking and rotational movement at the base of the hollow.

Drift: This is material deposited by a glacier and is made up of two main parts known as 'till' – deposited under the glacier – and 'outwash' which is formed by meltwater streams carrying particles of material from the debris under the glacier.

Drumlin: This is an oval-shaped hill formed from deposits within a glacier.

Erosion: This is the process by which rocks and landscapes are worn away by agents such as moving ice, wind, flowing water and sea/wave action.

Erratics: These are rocks or boulders which have been moved by ice sheets from their original location and left in other parts of the country during the ice age.

Eskers: These are long ridges of sand and gravel deposited by rivers which flowed under ice sheets.

Freeze thaw action: This happens when water trapped in cracks in rocks alternately freezes and thaws causing the rock to break up.

Frost shattering: This is similar to freeze thaw and is caused by water turning to ice and, as it melts, it expands putting pressure on the rock and eventually causing it to shatter.

Glacier: This is a large mass of moving ice which changes the shape of the land over which it is passing.

Grykes: These are deep joints or fissures on the top of limestone plateaus formed through the chemical reaction of rainwater and limestone. When deep enough they leave the surface area to form large blocks called **Clints** giving a 'pavement' like appearance.

Hanging valley: This is a valley which is situated on the slopes of a U-shaped valley and is a tributary of the main valley.

Mass movements: This is the general process by which rocks move under gravity and includes slumping, rockfalls, soil creep and mudflows.

Meanders: These are bends formed in the middle or lower courses of rivers.

Misfit stream: A U-shaped valley is usually occupied by a small river which is known as a 'misfit stream' since it was not the original river flowing through the valley.

Moraine: This is material which is deposited by glaciers. Different types include 'end-moraines'/'terminal moraines' formed at the front of the glacier as it melts, 'lateral moraines' formed at the sides of glaciers and 'medial moraines' formed in the middle of glaciers or at the edges of where two glaciers meet.

Plucking: This is the process by which moving ice tears rocks from the surface over which it moves.

River cliff: This is the steep bank of a river formed as the river undercuts or erodes the outside bank.

Stalactites/Stalagmites: These are formed underground in limestone areas where material in solution is deposited on the floor or drips from the ceiling of underground caves.

Till: Material deposited beneath a glacier (at the end of it) and consisting mainly of boulder clay.

Transportation: The process by which rock particles are carried by rivers or glaciers or wind.

Truncated spur: A piece of land, the bottom of which at one time jutted into a valley and was cut away or eroded by a glacier flowing through the valley.

U-shaped valley: A valley with very high steep sides and a wide flat bottom formed by a glacier flow through the original valley.

Weathering: The process by which rocks are worn away through physical action such as flowing water, wind or a chemical reaction between rocks and rainfall which may have become acidic.

Biosphere

The term Biosphere refers to the biotic response to specific climatic and other environmental conditions, such as relief and soils, which results in a variety of different types of vegetation. The various plants which exist on the earth's surface *inter-react* within a system called an ecosystem.

Soils – formation, properties and soil types

Soils are formed from the processes of weathering, erosion and deposition from parent rocks.

What is a soil profile?

Soils consist of several layers called horizons or soil profiles.

Figure 4.1 shows a basic, model horizon or soil profile.

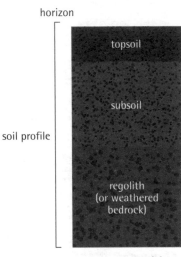

FIGURE 4.1: BASIC SOIL PROFILE

- The layer closest to the surface is called the Ao layer and contains the organic material derived from dead plants and other organisms. This is known as **humus**. Below the Ao horizon lies the A horizon which is the soil proper layer. This consists of a mixture of humus and other mineral particles.
- A further layer known as the B horizon is found beneath the A horizon consisting of coarser material. This is the sub-soil layer.
- There are two further layers called the C horizon consisting basically of weathered rock fragments, and the bottom layer or D horizon which is the parent rock layer.
- Depending on the variations within these different horizons, it has been possible to group soils into a classification system known as zonal soils.

How are soils formed?

Various processes are involved in the formation of soils in general. These processes depend on factors such as climate, relief, organisms, parent material and time.

- The first stage of the process involves the weathering of parent rock.
- The next stage of formation is that of adding water, gases, living organisms and decayed organic matter.
- Climate plays an important role in weathering of rocks. The quickest rate of breakdown occurs in hot, humid climates.

- If the rainfall is heavy the water moving downwards through the soil transports minerals downwards in a process called **leaching**.
- **Capillary action** is the process by which soil moisture moves upwards through the fine pores of the soil.
- Vegetation provides **humus** and more humus is found in tropical forests than in tundra areas.
- Two other processes are important, namely: (1) **Eluviation** which involves the washing out of material, i.e. the removal of minerals such as calcium and aluminium, and organic material from the A horizon. (2) **Illuviation** which is the deposition of this washed material in the sub-soil.
- **Organisms** within the soil affect the breakdown and decay of vegetation and therefore impact on the depth of the humus layer.
- Any living creatures such as insects and worms also affect the development of the soil.
- In areas with cool climates and where precipitation greatly exceeds evapotranspiration, a process known as **podzolisation** is common. Percolating rainwater becomes quite acidic as it passes through an acidic humus formed from falling pine cones and needles. This water dissolves and removes iron and aluminium oxides from the top-soil and leaves behind a high amount of silica in the A horizon which is bleached and drained of coloured minerals.
- When waterlogged conditions exist in the soil due to the loss of water from the soil being restricted (if for example the sub-soil is full of stagnating water which loses oxygen), a process known as **gleying** occurs.

You should be able to recognise three main soil types from their profiles. These include podzols, gleys and brown earth soils.

Top Tip

Make a list of the processes involved in soil formation and try to memorise it.

Podzols – soil properties

- These soils are found in a wide belt across the northern hemisphere, particularly in the areas of taiga or coniferous forests. Falling pine needles and cones create an acidic humus called **mor**. The soil has well-defined layers.
- The upper A horizon has an ash grey colour due to the removal of the minerals by leaching.
- With aluminium and iron oxides concentrating in the B horizon, a cementing effect takes place between the A and B horizons, forming a hardpan which seriously affects the drainage through the soil. This results in the upper layers becoming waterlogged.
- The B horizon is reddish brown in colour from the iron oxides.
- The sub-soil consists of weathered parent rock.
- Decomposition of the Ao horizon is very slow due to the cold climatic conditions and lack of soil biota.

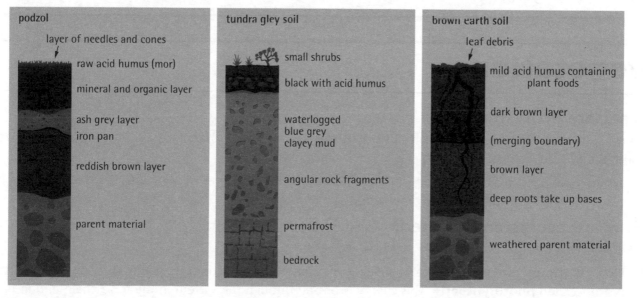

FIGURE 4.2: SELECTED SOIL PROFILES (PODZOLS, GLEY, BROWN EARTH)

Gley – soil properties

- The sub-soil in these soils remains frozen throughout the year.
- During the brief summer the ground surface thaws but the meltwater cannot drain freely due to the frozen sub-soil or permafrost layer.
- This results in the soil becoming waterlogged or **gleyed**.
- Bacterial action is very restricted due to the cold temperature.
- The A horizon contains black, acidic humus only partially decayed due to low temperatures.
- The B horizon is bluish grey in colour with clayey mud.
- Fragments of weathered parent material is often found within the B horizon.
- The cold conditions of the climate severely restrict the use of these soils.

Brown earth soils – soil properties

- These soils are associated with areas of deciduous forest.
- The humus layer is thick and generally fertile due to the variety of vegetation which is decayed.
- The humus is less acidic and is referred to as **mull**.
- Precipitation exceeds evaporation sufficiently to cause some leaching.
- The horizons merge more than the podzols due to the activity of earthworms and insects (biota).
- With the redeposition of iron and aluminium due to illuviation, the colour of the soil becomes increasingly **reddish brown**.
- Unlike podzols the soils tend to be free draining.
- There is a high clay content throughout the profile of brown earth.

How to analyse soil profiles

- Begin your analysis by noting the various constituents of the different horizons within the profile, beginning with the **top-soil**.
- Describe the thickness of this layer and refer to reasons for this, for example depth of humus layer and vegetation.
- Discuss each layer in turn.
- Refer to thickness, content, colour, texture, water content and whether the layers of the A, B and C horizons are **well defined** or whether **mixing** has occurred.
- Explain the properties by referring to factors such as climate, vegetation, processes of leaching, eluviation and illuviation and soil biota.
- From your analysis you should be able to deduce the type of soil which the given soil profile shows.

Vegetation

What are ecosystems?

An ecosystem consists of a **community of organisms** and their environment, such as tropical rainforests and hot and cold deserts.

Ecosystems can vary in size from very small pieces of land in the corner of a garden, to ponds or areas of bog or wetlands, to whole zones of desert areas or rainforest.

How are ecosystems formed?

All ecosystems depend on two basic processes, namely:

1. **Flow of energy**, the main source of which is the sun. This is absorbed by plants and converted by the process called **photosynthesis**. This energy is passed through the system in the food chain and in part of the chain depends on the preceding link for food and energy.

2. Recycling of nutrients, in which some nutrients are continually recycled within the ecosystem through plants absorbing them, animals and plants dying and decaying and returning nutrients to the soil.

Plant succession and climax vegetation

Certain plants may be affected by changing environmental conditions and this leads to a process of **community change** known as **plant succession**. The process of plant succession involves a series of stages during which the species of plant life changes. This whole process, from the initial establishment **(the pioneer stage)** of vegetation, to the final stage (or **climax**) is called a **sere**.

At the first stage the plants which colonise a completely new, perhaps bare, site such as a beach or sand dune, are called **pioneers**. **Lichens and mosses** present in the pioneer stage may be replaced by **grasses** which in turn may eventually give way to **shrubs and woodland**.

Climax vegetation is the final stage in the process of plant succession.

Climate is the principal factor in climax vegetation.

You may be asked to describe and explain plant succession to be found across a sand dune transect.

Top Tip

Always give the names of specific plants in your answers.

Marram grass

Sea couch grass

Ragwort

Sea holly

Sand sedge

Bog pimpernel

Lyme grass

Sand fescue

fixed sand

Marsh helleborine

sea

Sea rocket

Sea bindweed

Creeping willow

Sea spurge

Bird's foot trefoil

Sea sandwort

strandline   foredune   yellow dune   grey dune   dune slack

FIGURE 4.3: PLANT SUCCESSION ON DUNE BELT (PSAMMOSERE)

Your description and explanation should include points such as:

- In the **strandline** area, salt tolerant species such as sea rocket and saltwort can withstand the effects of sand and wind. These species have a high pH figure due to the high concentration of shell fragments.
- Dunes were formed from the sediments deposited along the sea shore which were blown inland. The first dunes to develop are known as **embryo dunes**.
- Pioneer species in these dunes grow side/lateral roots and underground stems (rhizomes) which bind the sand together.
- Some species which can tolerate the harsh conditions include **couch grass, marram grass and ragwort**.

Quick Test

From memory, draw the profiles of brown earth and podzol soils with labels.

- The embryo dunes change to **'yellow'** dunes, so called because of the yellow colour of the large amount of bare sand.
- **Mosses and lichens** begin to colonise the dunes.
- **Foredunes** contain species such as sea holly and sand sedge and have a lower salt content. Marram grass becomes a key plant in the build up of the dune.
- As the dunes begin to stabilise, **humus** accumulates from decaying marram grass and soil forms on the surface.
- Grey dunes and slacks contain sand fescue, bird's foot trefoil, heather and grey lichens. Due to the increase in organic content (humus), greater shelter and a damper soil a wider range of plants thrive here. **Slacks** are damp hollow areas between the dunes where water gathers.
- Plant life in slacks includes **marsh plants, rushes, alder and small willow trees**.
- In time, more mature vegetation forms on the older dunes, including woodland with **pine, holly and oaks** further inland.
- **Grassland** is found in areas with a high proportion of calcium carbonate from shell fragments whereas **heath** is more likely to develop on more acidic soils.
- This area is known as the **climax**.

Sample question, answer and comments

Question

Describe and give reasons for the changes in plant types likely to be observed across the transect as you move inland from the coast. You should refer to named plant species likely to be found growing at different sites and to influencing factors **such as** shelter, pH and distance from the sea.   16 marks

Answer (1) denotes correct point

At the strandline, the plants likely to be found are sea sandwort and sea rocket (1). These plant species are salt tolerant (1) and sand and wind resistant (1). They can also tolerate immersion in sea water (1). The high ph value (1) on the strand line is due to the high number of shell fragments (1).

At embryo dunes these plants are replaced by sea couch grass and lyme grass (1). These dune pioneer species grow side roots which bind the soil together (1). In the foredunes sea holly (1) and marram grass are the plant species found.

At the yellow dunes marram grass (1) dominates the dune as it can survive in extremely dry conditions. The humus content also increases on yellow dunes (1). As marram dies it is replaced by other grasses such as sand fescue (1). Due to shelter from sea and wind (1) more species are able to thrive (1).

Climax vegetation is the part in plant succession where no more big changes happen and the system and biomass is maximum (1). Trees such as pine and birch (1) are found here.

Quick Test

From memory write down the names of plants and trees which may be found in an example of plant succession.

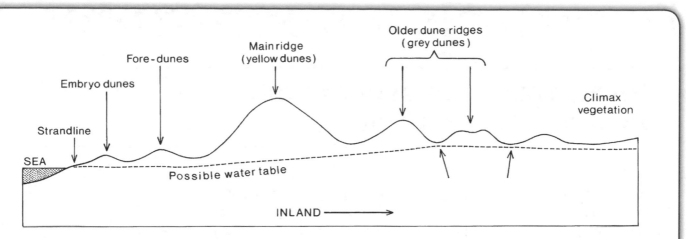

REFERENCE DIAGRAM Q2: TRANSECT ACROSS SAND DUNE COASTLINE

Comments

This is an excellent answer which gains full marks: **16/16**. The candidate correctly identifies the plant species at different stages. The candidate also gives appropriate explanations why these species are where they are. In answering questions such as these it is important to mention examples of species at each stage of the succession and to give some reasons why they occur.

Topic glossary

Capillary action: This involves the upward movement of water in a soil.

Climax: When a plant community reaches the mature stage of the succession process and becomes relatively stable and unchanging, it has reached the 'climax' stage.

Ecosystem: This is a community of plants which interact with each other.

Eluviation: This involves water percolating through the soil.

Gleying: This occurs where soils become waterlogged with stagnant water which is unable to drain away. The entry of oxygen into the soil is restricted turning the red iron oxides into blue grey iron oxide.

Illuviation: This is the process by which material is deposited in the sub-soil.

Leaching: This is the process through which minerals are carried through the soil by percolating water.

Parent rock: These are rocks from which particles are worn away to eventually become other rocks.

Population Geography

Demographic systems and population change

The following terms are important for you to understand and be able to use in your answers:

- **Birth rates.** This figure indicates the number of people per thousand of the population, born in any given year.
- **Death rates.** This indicates the number of people per thousand of the population who die in any given year.
- **Natural growth rate.** Subtracting death rates from birth rates gives a basic indication of how much the population is increasing each year per thousand of the population.
- **Average life expectancy.** This is a figure which indicates the average number of years a person can expect to live within any given country, e.g. male 67 years, female 70 years.
- **Infant mortality rate.** This rate indicates the number of deaths per thousand of the population in a year in any country of children under the age of one year.

You should be able to give reasons for any changes which may have taken place in populations in certain parts of the world.

Change happens because of:

- Variations in birth rates, death rates, migration patterns. These have a direct impact on the structure of the population of any given country.
- Factors such as wars, natural disasters and migration.

Censuses

- Most countries of the world have for some time recorded the number of people living within the country.
- The data is collected at regular intervals, for example every ten years in Britain, through a survey throughout the country.
- The census form contains a wide variety of questions relating to details such as date of birth, gender, employment status, ethnic origin and residence.
- The cost of carrying out regular and accurate censuses is very high.

Why do developing countries such as Pakistan or Malawi have more difficulties in carrying out a census than developed contries and why is the data is less reliable?

Reasons include:

- In countries where people move around a great deal, e.g. nomadic herdsmen, it is very difficult to keep track of population.
- If the standard of education is poor, the population may be unable to read and unable to complete the forms.
- Conflict such as wars and natural disasters, e.g. famine and floods, also make it extremely difficult to carry out an accurate census.
- Some governments may have political reasons for not having accurate census details, particularly if there are problems with ethnic minorities.
- Other reasons include, for example, the size of the country, difficulty with the terrain, e.g. mountainous areas, desert areas, rainforested areas.

Population structure

You are often asked in questions to describe and explain the population structures of Economically more developed (Developed) and Economically less developed (Developing) countries. How do you do this?

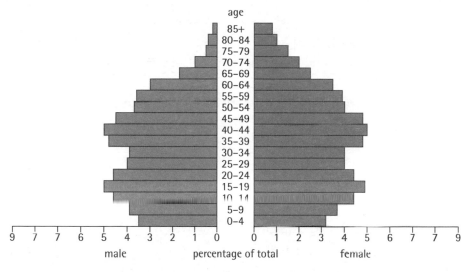

FIGURE 5.1 POPULATION STRUCTURE OF AN EMDC

When asked about the main features you should mention:

- A fairly low birth rate in both males and females.
- A bulge in the middle age groups, e.g. 15–60, which shows that most of the population are within these age groups.
- A fairly high percentage of the population within the upper age groups from 60+. This shows high life expectancy and population which has more old than young people.

What are the main reasons for this structure?

These include:

- The widespread use of artificial birth control.
- Changes in the status of women, with many women having jobs and careers in preference to marrying young and starting a family.
- The number of children per family is usually low due to widespread use of contraception and couples having children much later in marriage.

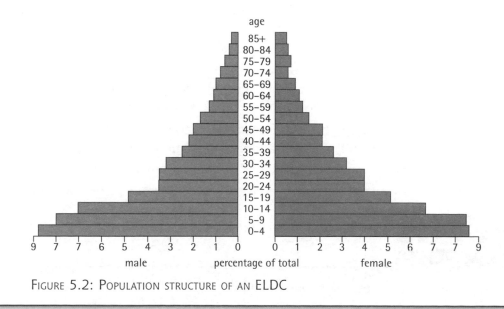

FIGURE 5.2: POPULATION STRUCTURE OF AN ELDC

The Human Environment

- The overall wealth of the country and the high standard of living enjoyed by its population.
- High standards of health care, education, housing, employment.
- Average income per head of population being high.
- People living longer due to high standards of health care.

This structure indicates a less developed country. Its main features include:

- A high birth rate in both males and females.
- A large proportion of the population between the ages of 0–15 years.
- A more definite pyramidal shape.
- The numbers of people in the upper age groups above 15 years decreases fairly rapidly.
- Very few people in age groups above 60 years.
- This means that average life expectancy is low.

This structure is due to:

- A lack of social and economic development of the country.
- Usually low standards of personal wealth, industrial growth, health, education, food supply, housing, health care and employment.
- The birth rate and death rates both being high.
- Families having a large number of children because infant mortality rates are so high.
- Children working and bringing in more income for the family.
- The children when grown can look after parents when they become elderly.

Top Tip

Make a list of the main differences in the diagrams you have just drawn and memorise it.

What consequences do different structures have for a country's economy and people? For a developed country, these include:

- Fewer young people, which deprives the country of a suitable workforce for future generations.
- More older people to be cared for due to increased life expectancy.
- Increased demand for more medical care and spending on care for elderly, e.g. retirement homes.
- A smaller group in the economically active age group which has to support an increasingly economically dependent age group through, for example, increased taxes to pay for health and social services.

For developing countries these include:

- **Overpopulation** which is said to exist whenever a reduction in the existing population would result in an improvement in the quality of life for the remaining population.
- A **lack of sufficient food** to meet the demand.
- **Inadequate housing** for the population, particularly in cities and towns.
- Vast numbers of people in less developed countries are forced to live in very poor accommodation such as **shanty towns**.
- These towns lack the **basic facilities** of sewerage, electricity and water supply.
- Consequently disease levels are high.
- **Unemployment** is high since there are far too many people for the jobs available.
- **Poverty** is widespread.
- Lack of services such as health centres, hospitals, doctors, sewerage systems, clean water supplies, schools and colleges, create problems of poor health and education standards.
- The percentage of the population which can read and write is usually quite low.

Quick Test

Draw a basic diagram showing the structure of a developed and a developing country.

In some questions you may be asked to compare different patterns of death by age group between developed and developing countries, suggesting reasons for the differences.

In your answer you should refer to:

- In developed countries the majority of deaths (80%) occur among the oldest age group – **65+** whereas this age group accounts for only **25%** of deaths in developing countries.
- In developing countries the majority of deaths occur in lower age groups, particularly young children.
- In the developed world this accounts for a very small proportion of deaths – **under 10%**.
- Deaths in the age group **16–64** are also much higher in developing countries than developed countries.

Why does this happen? Reasons include:

- Differences in life expectancy and infant mortality rates as a result of different living standards.
- Poorer quality of health and hygiene care available to citizens in less developed countries.
- Poor quality diet and food supply often leading to malnutrition and in extreme cases famine whereas in developed countries good quality food is widely available.
- Wide differences in levels of economic development with a lack of investment in housing, health, and general infrastructure in poorer countries.
- Low standards of education due to lack of schools and colleges and teaching staff because of lack of funding.
- Poor housing standards with shanty towns common to most cities in less developed areas.
- Widespread occurrence of infectious diseases due to poor sanitation, lack of medical drugs and health care as opposed to the high standard of health care in richer countries.

Demographic Transition Model

The graph in Figure 5.3 indicates changes in birth and death rates for the world in general over four separate periods or stages of time.

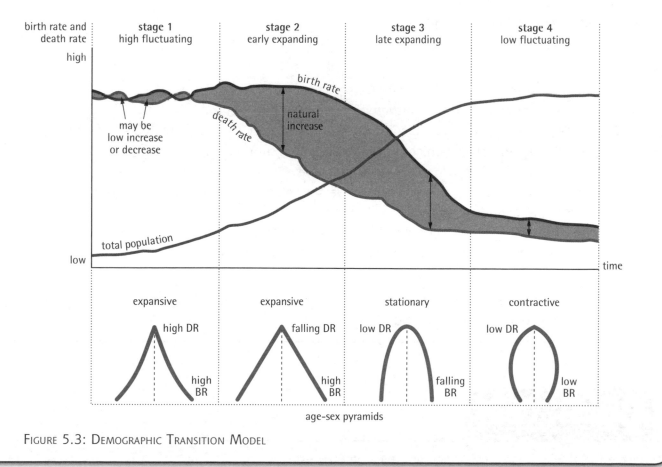

FIGURE 5.3: DEMOGRAPHIC TRANSITION MODEL

In questions you are asked to describe the pattern of birth and death rates in each stage and give reasons for these patterns. You should also, if given appropriate data, be able to suggest which stage a particular country might belong to.

Stage 1

- In this stage birth and death rates are both high, therefore growth is restricted.
- Most of Western Europe had this pattern about 250 years ago.
- Many people died at an early age from a wide variety of diseases, poor diets and poor living conditions.
- Large families were normal at this time.
- This pattern still occurs today in some of the least developed countries of the world.

Stage 2

- As countries in Europe became more industrialised, many people moved from the countryside to the industrial areas.
- Birth rates remained high, but as diets and medical care improved, the death rate gradually declined.
- Consequently the population grew rapidly.
- In Europe this happened during the period known as the Industrial Revolution.
- This process happened much later in other countries, particularly in Asia, South America and to a lesser extent Africa.
- During the second part of the twentieth century populations rose dramatically in these areas.
- Although droughts and famines still happen, there is now more food than ever before produced by farmers in less developed countries.
- Nutrition is much better due to the efforts of international aid agencies.
- Medicines which help control and prevent diseases, from simple measles to typhoid, are more readily available.
- Infant mortality and death rates are decreasing rapidly.
- Birth rates however still remain high.
- The result of this is rapid population increase in developing countries.

Stage 3

- In this stage the birth rate is beginning to decrease.
- The main reason for this is the introduction of methods of birth control.
- As diets, food supply and medical care improved, infant mortality rates decreased significantly.
- There was less need for larger families since more children survived.
- The process was repeated in many developing countries such as Mexico and India.
- A combined effort by the governments of many less developed countries and the work of international aid agencies led to a gradual reduction in the birth rates.
- Some countries, such as China, introduced laws to limit numbers of children to reduce the birth rate.
- Others tried to educate people on the benefits of using birth control measures.

Stage 4

- In this stage both birth and death rates are low and population growth is severely restricted.
- Most developed countries have reached this stage.

Quick Test

Make a list of why people live longer in developed countries.

- This is due mainly to the widespread use of birth control methods.
- Economic and social factors such as more employment opportunities being available to women, better education and health care standards have reduced birth rates.

Stage 5

- Some developed countries, such as the UK, France and Italy, experience this stage due to the birth rate dropping below the death rate.
- This gives an imbalance in the population structure where older age groups are larger than those of young people.
- Future workforce can be affected.
- Countries try to offset this by recruiting immigrant labour.
- This in turn can lead to civil unrest and problems of ethnic tension.

Although not shown on figure 5.3, it is possible to add a fifth stage to the original model.

Migration

Migration patterns

There are different types of migration, including:

Voluntary. This happens when people choose to move to another place.

Forced. This happens when the population has to move to another area against their wishes.

Long term. This occurs when people leave their home area and live elsewhere for long periods of time.

Short term. This happens when people move to an area for a short period of time, e.g. a season.

Rural and urban. This occurs when people leave the countryside to live in city areas.

By referring to 'push and pull' factors, you should be able to describe and explain migration patterns.

Rural and urban patterns

The factors which encourage people to leave the rural areas are termed **'rural push'** factors and those which attract people to the towns and cities are termed **'urban pull'** factors.

Rural push factors include:

- Lack of good farming conditions such as poor soils, dry climate, difficult terrain.
- Loss of farmland through, for example, land reform measures or inability to repay debts.
- Lack of employment through increased mechanisation.
- Low wages from agricultural employment and therefore low standard of living.
- Poor living conditions in terms of housing.
- Possible lack of educational opportunity.
- Lack of facilities such as entertainment, e.g. cinemas, theatres and other attractions especially for younger generations.
- Quality of health care may be severely limited through lack of hospitals and medical staff.

Urban pull factors include:

- Possibility of employment in a variety of jobs.
- Possibility of higher wages and therefore a higher standard of living.
- Offer of better housing conditions.
- Possibility of better educational opportunity for younger people.
- Possibility of family and friends already living in urban areas.

Barriers to migration

- Barriers to migration include legislation in receiving countries which limit the number of immigrants permitted to enter the country.
- Legislation may include references to marital status, ability to speak language of receiving country, professional or trade qualifications or political considerations or criminal records.
- Social barriers may include racial bias, problems with housing, education, employment and ethnic integration.

What advantages and disadvantages has migration brought to both the losing and receiving countries?

Advantages

- Receiving countries acquire labour for a variety of occupations, many of which may be less popular and difficult to fill from their own populations, or for which they have a skill shortage.
- The home countries of the immigrants benefit from money sent back and from skills that have been acquired when they return.
- There is also reduced pressure on resources while they are away.
- Incoming immigrants may add to the depth of culture of the existing population in terms of customs, language and traditions.
- If the receiving country's population is in decline, the immigrant population may help to reverse the trend.

Disadvantages

- Emigration creates problems for the areas which lose people.
- In some cases large numbers leaving results in depopulation (serious decrease in the numbers left).
- In small communities such as rural villages, e.g. in the Highlands of Scotland and villages in southern Italy and rural India, the loss of young people results in a highly imbalanced population structure.
- There is a large number of old people left. Often within a few years the community may eventually die out.
- New immigrants may increase the demand for housing, employment, education, health care and will obviously compete with the existing population for these services.
- In areas of relatively high unemployment additional competition for employment, especially if the new immigrants are willing to work for less money, creates great resentment.
- Demand for accommodation causes problems especially if the government is unable to supply the immigrants with housing.
- In developing countries the main result of this has been the emergence of shanty towns.
- In developed countries immigration from other countries can result in racial problems between ethnic communities especially where language differences exist.
- Racial discrimination is often common and shows itself in, for example, prejudice in employment and housing and in racial abuse.

You may wish to refer to actual countries which experience the advantages and disadvantages of migration. For example, out migration from Eastern European countries such as Poland, Lithuania, Romania and Hungary is high. This causes problems such as a lack of sufficient labour force in the losing country.

Many migrants have moved to Western European countries such as France, Germany and the United Kingdom in search of higher paid jobs. In some cases migrants have entered these countries illegally and often claim to be seeking asylum from persecution in their home country. This obviously can lead to the problems outlined above.

Geographical methods and techniques

You are often asked to interpret a variety of population graphs.

Interpretation of population data

- This skill involves being able to use population data to explain trends.
- Questions on these kinds of diagrams may ask you to examine the graphs, tables or maps and draw conclusions.
- For example you may be given a table showing various measurements of population for two or more countries.
- This may contain details on life expectancy, birth and death rates, infant mortality rates and other indicators such as medical provision.
- You may be asked to explain differences between the countries, e.g. life expectancy.
- Do this by describing the differences and then refer to reasons for varying levels of development between the countries.

Sample question, answer and comments

Question

Study Reference Diagram Q1 (Figure 5.3 on page 51) which shows the four stages in the Demographic Transition Model.

Referring to a country or countries which you have studied, describe and account for the changes in population from the beginning of stage 2 to the end of stage 3.   18 marks

Answer (1) denotes correct point

There are four stages to the Demographic Transition model but the two stages I am interested in are the Early expanding and Late expanding stage which are 2 and 3.

In many countries in Africa like Ethiopia, Mali, Somalia etc. they are going through these stages at this moment in time.

Stage 2 is the Early expanding stage where after a high birth and death rate in stage one the death rate drops dramatically (1). This is due to the introduction of many things. One of which is health care (1), as before many people were dying of diseases which could be cured with simple jabs (1) like whooping cough (1) so less people die from the disease. The water sanitation has also been improved (1) as people begin to separate waste from clean water and by doing this diseases are reduced (1) as people now have clean water available. People's standard of life has improved but the birth rate remains high (1) as people still produce babies as they did earlier so their family live on and at this stage the population dramatically increases (1) due to a lot more people dying than being born.

In stage three the death rate is low (1) and the birth rate drops (1) as people in countries like Ethiopia now realise that life is better and less babies die at a young age so they begin to produce less babies as this would mean that they would not be able to afford food if the babies were still being born (1).

The introduction of family planning and providing the families with condoms and other contraceptive methods (1) come into play. The population at this point is still increasing (1) but beginning to level off as birth rate reduces to alongside death rate.

Comments

This is basically a good, sound answer covering the essential points in both parts of the question. However the answer is spoiled by several unnecessary additional statements such as the introductory sentence. There is also a certain amount of carelessness in parts of the answer in, for example, part two when referring to Ethiopia as an example of a stage 3 country. The last sentence in part one referring to 'more people dying than being born' is again confused and careless.

Throughout the answer there are many relevant points such as good examples of stage 2 countries, death rate dropping dramatically, reference to health care and people dying from simple diseases. The comments on water sanitation reducing disease and the fact that there is an increase in population is credited.

In the comments on stage three, the candidate correctly refers to low death rate and a falling birth rate and the link between quality of life and number of babies born. Comments on family planning and contraception are also worth additional marks as are the final statements on the growth rate levelling off due to parity between birth and death rates.

There are sufficient points made to gain a total of **13 marks out of 18 marks**.

Top Tip

In answering questions such as these, avoid giving a simple list of changes. For each change mentioned write at least one sentence of explanation.

Topic glossary

Active population: That section of the population of a country which is economically active and working.

Birth rate: The number of births per thousand of the population in any country in any year.

Census: A numerical count of the population financed and carried out by the government at set periods of time, e.g. ten year intervals.

Death rate: The number of deaths per thousand of the population of any country in any given year.

Developed countries: Sometimes referred to as 'more economically developed countries' include countries which have a high standard of living or high physical quality of life.

Developing countries: Often referred to as 'less economically developed countries' in which the population generally has a low standard of living.

Empty lands: Areas of the world which have a low population density, e.g. mountains and deserts.

Environmental factors: Factors such as climate, relief, soil and water supply which can influence the distribution of population in an area.

Gross Domestic Product (GDP): The value of all goods and services of a country produced in one year; used as an indicator of the wealth of a country.

Infant mortality: The number of children below the age of one year who die per thousand of the population.

Life expectancy: The average age a person can expect to live in any given country.

Population density: The average number of people within a given area, e.g. 100 per square kilometre.

Population Growth Model: This shows different stages of population growth based on the relationship between birth and death rates.

Population structure: This refers to the grouping of the population of a country by age and sex.

Standard of living: This is the level of economic well being of people in a country.

Rural Geography

Agricultural systems

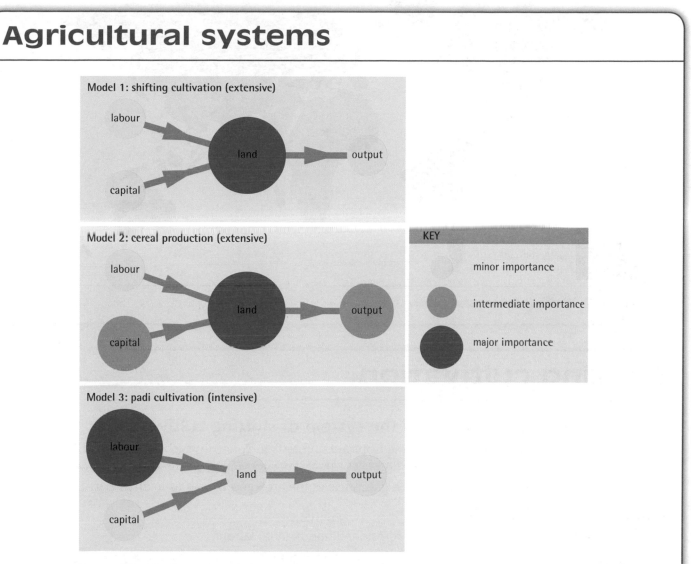

FIGURE 6.1: MODELS OF AGRICULTURAL SYSTEMS

Any system involves **inputs, outputs and processes**.

- In farming inputs include **physical factors** such as climate, landscape and soils. **Human factors** include population, land ownership and cultural background, and **economic factors** such as capital investment, technology, government influence, transport and markets.
- Outputs include the crops, livestock and farm produce.
- Certain processes occur between the input and output stages, for example ploughing, sowing seeds, cultivation, harvesting and transporting crops.

Which systems should you know about?

For the purposes of the examination, you need to know about three particular farming systems, namely, **shifting cultivation, intensive peasant farming and extensive commercial farming**.

Top Tip

Memorise at least three inputs and three outputs of each system of farming.

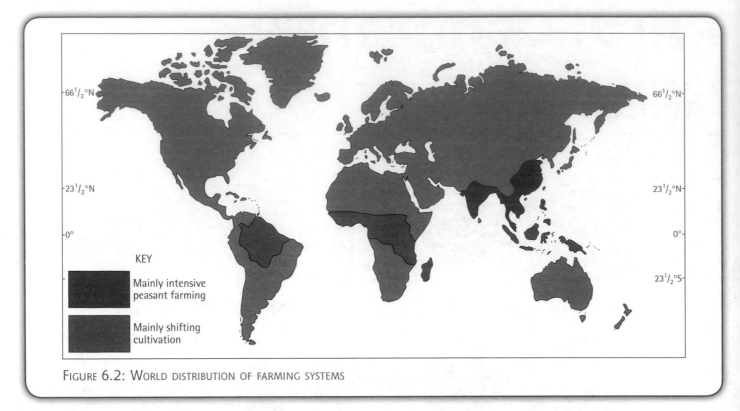

FIGURE 6.2: WORLD DISTRIBUTION OF FARMING SYSTEMS

Shifting cultivation

What are the main features of the system of shifting cultivation?

This system of farming is found mainly in the tropical rainforests in countries such as Brazil and Nigeria. The system involves a group of people clearing an area in the rainforest, and growing various crops until the soil becomes exhausted. When this happens, the group leaves the area and moves to another part of the forest.

Inputs

Inputs consist basically of areas of cleared forest land and manual labour.

Processes

- The natural forest is cleared either by cutting down the trees or by burning them.
- A fairly large area is cleared for cultivation. Other areas are left fallow to maintain fertility.
- Ash from the burned trees provides some fertiliser.
- Crops are planted in the ash covered ground.
- As the soil becomes less fertile, other crops which can grow in less fertile soils are grown.
- Other processes involve harvesting.

Outputs

- Outputs may include crops such as yams and manioc, fruit trees, coca beans and sometimes tobacco.
- As the yield decreases over the years and the ground becomes more and more infertile, the whole site is abandoned.
- The whole system is geared basically to supply the food requirements of the tribe.
- Diet can be supplemented by food from hunting, fishing or gathering fruits and plants.
- Although primitive, the whole process is very well suited to areas which are environmentally difficult, e.g. poor soils.
- Shifting cultivation can be environmentally friendly in that, given time, the forest can recover.
- Deforestation is only temporary.

Does the landscape of this system have any particular features?

- Temporary shelter is provided by huts and only parts of the area are actually under cultivation at any given time.
- The whole area may use as much as 20 acres.
- A total area of up to 100 acres may be need to support a group of about 20 to 25 people.
- Different parts are farmed at different times over a period of perhaps up to ten years.
- Population density is low.
- Tribes are usually quite small and widely scattered.

Have changes affected the system?

- Other land uses have affected shifting cultivation.
- These include mining, ranching, settlement, roads, plantations and hydro electric schemes.
- The shifting cultivators have been forced to move further and deeper into the forest.
- In some cases the tribes are forced to abandon the system altogether and seek an alternative form of existence.
- In some areas the tribes have been forced to move into reservations.
- Their traditional way of life is therefore threatened by coming into contact with modern civilisation.
- Natives have been badly affected by their lack of immunity to western diseases.
- Some of the tribes seek work on the plantations, others in forestry.
- Despite these changes, up to 150 million people still remain in the system of shifting cultivation throughout the world.

Intensive peasant farming

This type of farming is found throughout South East Asia, including India.

Several different types of crops may be grown depending on climatic conditions. In wetter areas rice is the main crop. There are several different varieties of rice which can be grown but the most important is wet rice.

Since the growing season is comparatively short (100 days at temperatures above 20 degrees C), two or three crops can be grown in a single year.

Inputs

- Inputs include physical factors such as flat land or, where this is limited, terraces on slopes, impervious soils and sufficient water to flood the fields in which the rice is grown.
- These are known as rice padi fields.
- Natural fertiliser may be added after the seeds are planted.

Processes

- If the fields are flooded, the floods may deposit silt which adds to the fertility of the soil.
- Where rainfall is deficient, primitive irrigation is used to bring additional water to the fields.
- Seeds grown earlier in nursery beds are transplanted to the fields which are under several centimetres of water.
- Stone walls are built around fields to retain water.
- The system is highly labour intensive throughout all stages.
- To help maintain fertility of the soil, rice stubble, algae and animal manure are ploughed into the fields.
- Soil erosion is prevented by the building of embankments to protect fields or terraces to prevent soil creep.

The Human Environment

- Machinery is hardly used in any of these stages.
- Livestock such as cattle are used as draught animals to pull ploughs and carts.
- The fields are allowed to dry out when the crop matures to allow the rice to ripen.
- Being a labour intensive system, **population density** is usually very high in most areas.
- The **land tenure system**, whereby farmers have inherited **small** plots of land from their fathers, makes the use of machinery difficult.
- Large families mean children can be used as extra labour at no extra cost.

Top Tip

Look at the landscape in figure 6.3. Try to describe in your own words the main features and try to remember your description.

Outputs

- Rice is the main crop grown in the wetter areas.
- The most important variety is the 'wet' rice grown in flooded fields.
- Usually two crops are grown each year on the best land.
- Other crops include wheat and maize.
- In drier areas crops grown include cereal crops such as wheat, millet and sorghum, chick peas and, in some areas, manioc.
- Manure is provided by cattle for fertiliser.

What does the landscape look like in areas with this system?

The main features include:

- Small fields due to the land tenure system.
- Rice planted under water.
- A distinct lack of mechanisation and a high number of workers working in the fields planting and harvesting crops by hand.
- Irrigation ditches dug to transfer water to fields.
- The use of animals such as oxen to draw carts and transport crops.
- Water is retained in fields with the help of embankments.
- Slopes are terraced to increase the land available and to conserve soil and water content.

FIGURE 6.3: TYPICAL PEASANT FARMING LANDSCAPE

What changes have affected crop production and what part has the Green Revolution played in these changes?

Governments of countries with peasant farming have tried to introduce modern methods and other measures to change the system.

These measures have been described as the **Green Revolution**, the main features of which are summarised in Figure 6.4

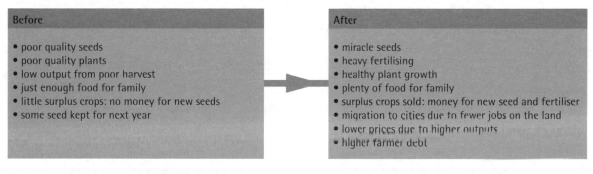

Before	After
• poor quality seeds • poor quality plants • low output from poor harvest • just enough food for family • little surplus crops: no money for new seeds • some seed kept for next year	• miracle seeds • heavy fertilising • healthy plant growth • plenty of food for family • surplus crops sold: money for new seed and fertiliser • migration to cities due to fewer jobs on the land • lower prices due to higher outputs • higher farmer debt

FIGURE 6.4: GREEN REVOLUTION

The Indian government introduced various schemes to improve farming during a programme of five and seven year plans which involved a range of measures, including:

- Land reform schemes whereby small farms resulting from the land inheritance system have been **amalgamated** into larger farms.
- Lending money to farmers to improve their farms.
- Introduction of **miracle** seeds in order to increase yields.
- Encouraging the use of chemical fertilisers to improve soil fertility.
- **Increasing mechanisation** by using tractors and other farm machinery.
- Employing **agricultural advisers** who provide training schemes for farmers.
- Spraying **insecticides** onto crops to destroy insects which damage crops.
- Introducing **modern irrigation** methods to replace inefficient methods such as inundation canals.
- Raising the level of technology used on farms such as motorised ploughs.
- Introducing laws to reform the inheritance system.

What effects have these changes had on people and the landscape and have they been successful?

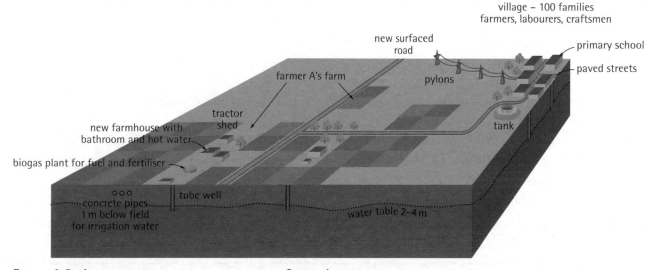

FIGURE 6.5: A MODERN AGRICULTURAL LANDSCAPE IN SOUTH ASIA

- All of these measures were designed to increase yield and output from the farms and the changes have had some success.
- For example, in India during the 1970s and 1980s, production doubled during these decades because of the Green Revolution.
- The changes also brought problems because using fertilisers, pesticides and other new techniques cost farmers a great deal of money.
- Farmers were encouraged to borrow money from banks.
- As yields increased prices fell due to increased supply.
- Many farmers did not earn enough money to repay the money they had borrowed and lost their farms.
- Only the richer farmers benefitted.
- Mechanisation left many farmers unemployed.
- Many farmers migrated to other areas or large cities.
- Land reform often deprived many farmers of the opportunity to own their own farm.
- The Green Revolution has been partially successful in some areas, but in other areas the success rate has been much less.

Commercial arable farming

The main aim of this system of farming is to produce outputs which are sold in the domestic and the export markets, making large profits. The system is found in countries such as Canada, the USA, Australia, Russia and the UK.

Inputs

Physical inputs include:

- **Landscape.** Land is generally flat. This allows mechanisation in ploughing and harvesting and transportation of the produce.
- **Soils.** Soils are often rich in humus, deep and very fertile such as the black earth soils of the Prairies and the Steppes of Ukraine.
- **Climate.** Summers are usually warm and precipitation is high enough.

Human inputs include:

- **Capital** in the form of:
  - **Land.** An extensive acreage of land sometimes in excess of several thousand square kilometres.
  - **Mechanisation.** Extensive grain farms make use of the most modern of machinery including combine harvesters, ploughs and aircraft to spray crops with insecticides.
  - **Fertilisers/pesticides.** Vast sums of money are spent on artificial fertilisers and pesticides to increase yield.
  - **Transport.** The system depends on the availability of a highly efficient transport (rail/water) network for the distribution of the produce.
  - **Improved seeds.** Seeds which can produce higher yields are used extensively.
  - **Selective irrigation.** Some areas need to supplement rainfall with irrigation schemes.
  - **Labour.** Labour input into extensive commercial farming is low due to mechanisation.

Output per person is relatively high.

- **Population density** in areas with this type of agriculture is quite low due to mechanisation meaning less need for labour.

- **Political.** Governments provide grants, subsidies and negotiate trading alliances with other countries such as those in the EU.
- Tariffs are agreed to protect farmers from foreign competition.
- Colleges and agricultural training and research centres are provided.

Processes/Methods

Processes involved in commercial arable farming include:

- Ploughing, fertilising, planting, harvesting, storage, transportation, maintenance of machinery.
- Strip farming (ploughing land at right angles to the wind) with wheat and grass grown in alternate strips is becoming more common.
- Farmers often hire squads of specialist workers with combine harvesters rather than harvest their own crops.
- Crop farms invest heavily in a variety of buildings for livestock and storage purposes.
- A great deal of money is also invested in protection of livestock and produce from disease, e.g. insecticides and veterinary costs.
- Other processing costs involve construction of processing plants, advertising and marketing and continual research into improving farming methods and techniques.
- Produce is transported mainly by rail to ports and other cities.

Outputs

- Output consists mainly of spring wheat and cattle in many areas, although other crops include barley and sugar beet.
- Winter wheat is grown in warmer areas.
- Fodder crops such as alfalfa are also grown for large cattle ranches.

Landscape

FIGURE 6.6: TYPICAL FIELD PATTERN – COMMERCIAL GRAIN FARMING

- Figure 6.6 shows the field patterns and landscape typical of a large-scale commercial farming area and is based on a wheat farming area in the American Plain lands.
- Farms growing cereal crops tend to have large, very regular field patterns.
- Farms have been amalgamated in large units often in excess of 400 hectares.
- In the Midwestern states of the USA fields are laid out in large rectangular blocks measuring sometimes between 100 and 300 acres.

What changes have affected commercial arable farming and have they been effective?

- Capital input has continually increased especially in terms of mechanisation.
- More recently farmers operate within a cooperative or alternatively hire companies which supply labour.
- Hiring machinery and labour is more cost effective for farmers.
- The introduction of more mechanisation has resulted in more reductions in labour input and reduced population.
- Crop yields have increased by using seeds which are disease resistant and have faster growing properties.
- Yields have also been increased through the use of increased amounts of fertilisers and pesticides.
- Introducing genetically modified crops has been a further change although this is highly controversial.
- Measures to improve road and railway services are constantly being reviewed.
- **Diversifying** by introducing other crops is a further change.
- The natural environments of areas with this type of farming can also be affected by these changes, e.g. increased use of fertilisers can cause rivers to become polluted. Often the farms depend greatly on river management projects for water used in irrigation. These can affect the environment through valleys being flooded, destroying natural vegetation and the habitats of local wildlife.

Do these changes cause problems?

- Smaller farmers find it more and more difficult to compete with larger enterprises in terms of increasing capital input.
- Smaller farmers have therefore had to sell their farms to their larger competitors.
- Many farming employees have lost their jobs due to increased mechanisation.
- Population density has significantly decreased.

Has this type of farming affected settlement patterns?

- Settlements in these areas have a recognisable pattern of regular spacing.
- There is a hierarchy of settlements ranging from hamlets to towns with fewer larger settlements providing services for the larger number of smaller settlements.
- This spacing is due partly to the field patterns, the road and railway network and the service function of the settlements.
- Regular spacing of the service centres ensures that each centre maximises its share of the available consumers in the area.

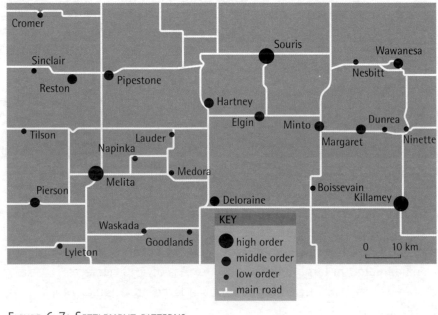

Figure 6.7: Settlement patterns

Geographical methods and techniques

You may be asked to describe and analyse land use data and crop yields shown in maps, diagrams and tables.

- If you are **explaining differences** between farms or areas in terms of output figures you should refer to the main points of comparison, e.g. similarities in main crops grown, or differences between them.
- You should quote directly from the figures given.
- If comparing land use maps, note the changes in land use and the size of fields.
- If asked to explain the changes you can refer to factors affecting the **farmer's decisions on what to farm**.
- Usually the main factors producing **change** will be economic such as capital investment in new machinery, fertiliser, seeds, etc., and profit from farm produce.
- Political factors would include reference to government policies and trade alliances such as the EU.

Sample question, answer and comments

Question

(a) Describe three main features of either extensive commercial farming or intensive peasant farming.   6 marks

(b) Study Reference Table Q1A.

Reference Table Q1A (Agricultural change in the Punjab)

	1960	1970
Amount of land irrigated by tube wells (%)	35	70
Fertiliser application (kg/hectare)	20	271
Area under high yielding varieties of wheat (hectares)	69	169 972
Wheat yields (kg/hectares)	1152	367

The Green Revolution was an example of change in areas of intensive peasant farming.

For either the Punjab or another area of intensive peasant farming which you have studied,

(i) describe the changes which have taken place, and   6 marks

(ii) comment on the successes and failures of these changes.   6 marks

Answer (1) denotes correct point

(a) *Intensive peasant farming has three main features as follows:*

It is usually carried out on small areas of land (1) very intensively farmed (1). There are many workers (1) for each area of land and very little or no machinery is used (1).

It usually occurs in developing countries and often involves only one crop, e.g. rice (1) in India.(1).

Comments

This answer contains sufficient detail to obtain **6 marks**. These are gained for points relating to small areas, labour input per area, lack of machinery, location in developing countries and monoculture, i.e. one crop grown.

The answer obtains a total of **6 out of 6**.

Answer

(b) (i) *In India the Green Revolution has had various degrees of success.*

Land which was previously 'fragmented' has been joined together (1) and landless workers have been given sections of land (1). The land owners land is no longer fragmented but is all in one area (1)(R).

Machinery has been introduced (1) and is used in areas, with new irrigation methods (1). Fertilisers and pesticides have been introduced (1) and their uses have been shown to be successful (1).

Farmers have been educated (1) in the use of these new materials.

New high yielding varieties (HYV) (1) have been introduced to replace the old seeds.

Comments

The answer contains some comments which are not awarded any marks, such as the first sentence. Marks are achieved for the statements on 'joining of previously fragmented land, workers given sections of land, machinery being introduced, new irrigation methods, fertilisers and pesticides, education of farmers and the introduction of new, high yielding varieties of seed'. The candidate sometimes repeat points (as indicated by *(R)*) and would gain no extra credit for this

The answer has therefore been good enough to obtain a total of **9 marks out of 6 marks**. However **only 6 marks** are available for this part of the question, therefore the total marks gained is **6 out of 6**.

Answer

(b) (ii) *The changes have been successful in some ways and unsuccessful in others. The new HYVs have been replaced by even higher yielding varieties (1).*

The problem has been that only the already better off farmers have been able to afford the new HYVs (1) and machinery and irrigation methods and have therefore increased their yields (1) but the already poor farmers have been unable to afford (1) these new things and therefore have become poorer because they cannot compete with the competitive prices (1) the better off farmers provide. Or they have taken out loans (1) with extortionate high interest rates (1) and have been unable to repay them and ended up further in debt (1).

Comments

The answer contains more than sufficient points to earn **8 marks**. Points are gained for comments on 'higher yield varieties being introduced, only affordable by better off farmers as well as machinery/ irrigation, increasing their yields, offered at competitive prices which poor farmers cannot compete with, resulting in loans which cannot be repaid causing further debt'. This answer therefore obtains **6 marks out of 6**.

Overall the total marks for parts a and b amount to **18 out of 18**.

Topic glossary

Agribusiness: The operation of a large-scale farm which resembles a factory with large investments in farm property, maintenance, machinery and technology.

Arable farming: Farms where the main activity and income source is the growing of crops.

Casual workers: Workers employed at specific times during the year, for example at harvesting.

Cereal crops: Crops which are grain crops such as oats, barley or wheat.

Common Agricultural Policy: A system used by the European Union to give farmers guaranteed prices for their products.

Crofting: A type of mixed farming found in northern Scotland which is not very profitable and farmers often have to supplement their income by doing other part time jobs.

Crop rotation: A system designed to maintain the fertility of soil by growing crops in different fields from time to time.

Diversification: Adding different enterprises to the farm in order to improve income and allow the farmer to become less dependent on income from farm produce.

Drainage: If the underlying rock is clay, bogland and marshland may develop.

Farm system: The relationship between inputs, processes and outputs on a farm.

Fertilisers: Substances which may be organic or chemical and are added to soil to increase fertility and improve crop yield.

Inputs: The basic needs of a farm before the farmer can begin to farm the land such as seeds, livestock and machinery.

Insecticides: These may be sprayed on to crops to kill insects which may be attacking crops and therefore destroying the yield.

Outputs: The end product from the inputs and the processes of production on the farm.

Pasture: Land grazed by livestock. Some of this may be permanent, some temporary and some might be poor pasture which has been improved by, for example, underground drainage schemes.

Processes: This is the work done on the farm and obviously varies according to the type of farm.

Quotas. These are limits imposed on farmers in order to limit the output of certain types of produce to avoid surpluses and therefore a drop in prices.

Top Tip

Read the glossary and for one letter A to D name a feature of farming and state in your own words what this feature is.

Industrial Geography

Industrial systems

Industry can be defined as the work which is done for economic gain. To help our understanding of this complex process, industry is usually classified into several different categories.

Primary industry: These are industries which **extract** raw materials directly from the earth and include farming, forestry, quarrying, and mining.

Secondary industry: These industries **manufacture** products.

Tertiary industry: These provide a service for consumers, including, e.g. retailing and wholesaling, transport and administration.

Quaternary industries: These are based on technological development, e.g. information technology, scientific research and development.

What are the main features of an industrial system?

All categories of industry operate as systems based on **inputs, processes and outputs**. These are summarised in Figure 7.1.

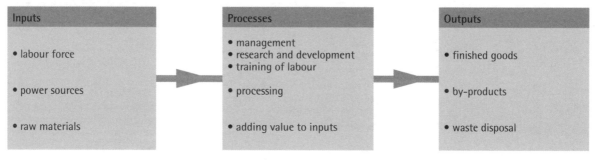

FIGURE 7.1: INDUSTRIAL SYSTEM

The chosen location of any industry is influenced by a variety of factors.

Figure 7.2 summarises these factors.

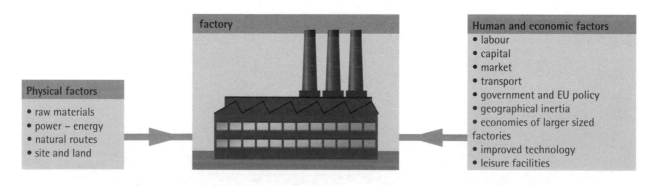

FIGURE 7.2: FACTORS INFLUENCING THE LOCATION OF INDUSTRY

When answering a question on this topic you should refer to the following:

Raw materials

- These are the materials from which products are made. Some industries e.g. mining and farming have to be located at the source of these materials. Others may have to be located very close to them so as to reduce the cost and difficulties of transport.

Labour supply

- All industries need labour. Modern industries, however, need to consider where labour is available and perhaps locate close to it.

Power

- In the past, power was provided by either water or steam which meant that industries had to be very close to either or both of these sources. River sites or coal fields were often chosen as suitable locations.

Transport

- Good access to industries is necessary both for raw materials, transporting the finished product and for labour supply.
- Many industrial sites are located near major roads or motorways, railway lines, in river valleys or at the coast, or near major airports.

Market

- The market may be very local as in the case of retailing industries, or can be worldwide as in the case of multinational companies.

Government

- Governments often offer a range of incentives to attract industry to a specific area or site, e.g. to ease the unemployment rate.

Industrial/geographical inertia

- When a large industry has grown it may remain in an area for cost reasons although the original location factors no longer apply. When this factor applies it is known as industrial/geographical inertia.

Economies of larger factories/improved technology

- Large factories often locate beside each other in order to share the benefits of cost of land or reduce transport costs if trading with each other.

What factors attract foreign manufacturing industries to an area?

At national level government may act to attract foreign industries to different parts of the country.

Measures employed include:

- Offering incentives to firms to go to specified areas.
- Offering financial assistance with labour costs.
- Offering tax incentives to foreign companies.
- Working closely with local government to reduce development costs.
- Offering help with the cost of improving local infrastructure.
- Offering financial assistance to older industries during periods of recession.

Reasons for adopting these measures include:

- Boosting the economies of depressed areas.
- Helping to reduce the unemployment rates.
- Helping with re-industrialisation of older industrial areas.

Quick Test

From memory draw a star diagram to show the main factors affecting industrial location.

- Helping to decentralise industry from economically stronger areas to weaker areas.
- Encouraging foreign investment from non-European countries, for example Japanese and American companies, in areas such as the Midlands and South Wales in Britain and in Eire. This allows these countries to trade within the European Union.

Industrial landscapes

Using a given OS map extract you may be asked to describe the main features of an industrial landscape and explain the main location factors. How do you do this?

The first task is to correctly identify the type of industries present, whether they are old traditional or new modern industries.

Figure 7.3 illustrates three different industrial areas which can be seen on an O.S. map.

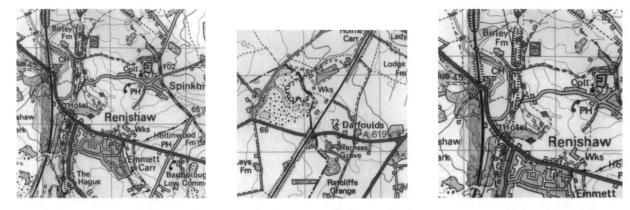

FIGURE 7.3 (A, B, C): SELECTED INDUSTRIAL SITES ON AN ORDNANCE SURVEY MAP

Site 1

- Primary industry, including mining, quarrying, forestry estates and farming, should be easily recognised on OS maps. Their location is due to the presence of the raw material being extracted.
- If you are asked about former industries, look for 'dis' meaning 'disused'. This indicates old mines or quarries and is a guide to industrial change.

Site 2

- Large manufacturing industries are easily spotted by the shape of the buildings on the map as shown on Figure 7.3 and the word 'works'.
- Power stations are also easily identified by written descriptions beside them.
- Similarly 'mills' and 'distilleries' will be described.
- Most older industries in a city will be located close to the centre of the city, perhaps surrounding the central business area.
- Most will be near the main road and rail arteries leading to the centre.
- Many industries may be located along one or both banks of a river.
- Newer industrial units are often found on the sites of closed older industries.

Site 3

- These are found in industrial estates situated within or on the outskirts of large urban areas. The words 'ind. est.' indicate their presence.

Top Tip

Try to memorise three map features which indicate the presence of 'primary', 'secondary' and 'new modern industry' on an OS map.

- Buildings will be smaller and laid out in a more planned manner than the larger factories.
- There may be easy access to a nearby motorway.
- 'Science parks' and 'enterprise zones/trading estates' have emerged in recent years and will be named on the map.

Industrial change

You could be asked to describe and account for industrial change in an area you have studied.

Throughout the UK many older industries have declined for several reasons including:

- **Exhaustion** of raw materials.
- **Loss of markets** for products such as ships, iron and steel and textiles.
- **Increased competition** from other areas, e.g. South Korea.
- **Use of new technology** which older industries have failed to develop.
- **Withdrawal** of financial assistance due to changes in government policies.

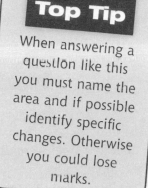

Top Tip

When answering a question like this you must name the area and if possible identify specific changes. Otherwise you could lose marks.

Case study of changing industrial area – South Wales

You could use South Wales as a case study. The changes which happened in this area are summarised in Figures 7.4 and 7.5.

Main changes include:

- The first change to the area took place about 200 years ago.
- At this time the area changed from a farming economy to an industrial area with the emergence of heavy industries such as coal mining and iron and steel factories.
- The area had the main raw materials needed for the growth of these industries, namely coal, iron ore and limestone.

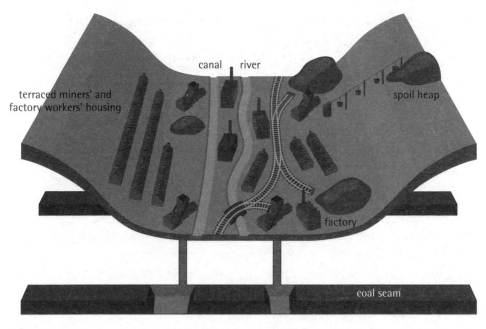

FIGURE 7.4: INDUSTRIAL CHANGE: SOUTH WALES IN 1910

The Human Environment

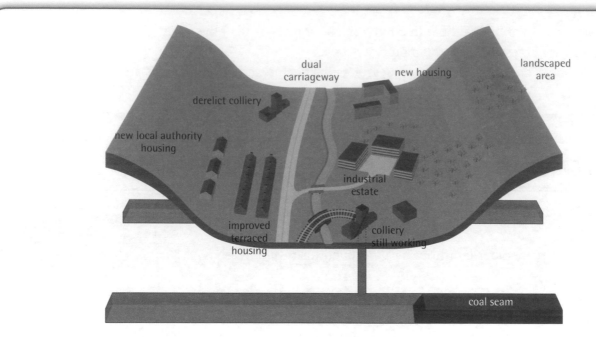

FIGURE 7.5: INDUSTRIAL CHANGE: SOUTH WALES IN 1985

- Coal and steel were also exported abroad to many of the countries of the British Empire including, for example, India.
- The three main factors responsible for the growth of industry in this area were:
  1. availability of raw materials, 2. the demand from a home and foreign market, and 3. a plentiful supply of labour.
- Change began to occur during the 1950s.
- The raw materials which were once in plentiful supply began to run out.
- The most important blow, however, was the decline in the markets for traditional products.
- Mines and steelworks closed at an alarming rate, leading to high unemployment.
- This led to many unemployed people moving away to work elsewhere.

Recent changes in South Wales include:

- Several new modern factories were encouraged with government aid during the 1970s to set up production in the area in places such as Bridgend.
- These industries included electronics companies, sweet making factories, shoe making and a variety of other modern industries.
- The original location factors were changed. New factors included:

Transport

- New industrial estates were built next to motorways such as the M4. This reduced transport costs and improved the efficiency of the system.
- These transport links were also essential for the labour force.

Government

- British and foreign firms were attracted to the area of South Wales through government financial incentives, such as grants for new machinery and retraining schemes.
- As a result, a fairly wide variety of different manufacturers were attracted to the area including shoe manufacturers, TV manufacturers and confectionery companies.
- Old closed mines were reopened as tourist and educational centres.
- In the Rhondda valley, mining villages were rebuilt as well as new theme parks, boating areas and dry ski slopes.

How did these changes affect the area?

- Old traditional factories were demolished and replaced by new modern industrial estates.
- For a time the landscape was scarred by old buildings which had fallen into disrepair and were an eyesore. These were eventually demolished.
- Modern buildings replaced old dilapidated buildings.
- New road systems replaced older roads making transport much easier.
- New modern planned industrial estates replaced old worn out factory buildings.
- The modern industrial environment is now relatively **pollution free** with several greenfield sites home to new modern light industrial units.
- The whole area became a cleaner, more environmentally friendly and pleasanter place in which to live and work.
- Closing older industries had social and economic effects on the area.
- Many workers became unemployed and moved to other parts of the country in search of work.
- Unemployment had an economic impact on local businesses which lost custom.
- Some of these problems were addressed by the development of new industries which offered employment to those who lost jobs in older industries.

Geographical methods and techniques

You should know how to analyse various types of industrial data.

- For example, you may be given data on employment figures for specific areas over a period of time.
- You should begin your analysis by drawing comparisons between the figures noting the main **similarities and differences**.
- Use your knowledge of the location factors to make some comment on the trends.
- A similar question may ask you to compare changes based on data relating to the number and types of firm in an area between specific dates.
- Note the main changes, particularly in terms of the types of industry.

Sample question, answer and comments

Question

Study Reference Diagram Q1. (See Figure 7.2, page 68.)

With reference to a named industrial concentration in the European Union which you have studied,

(a) explain why physical factors led to the growth of early industry, and   8 marks

(b) suggest why human and economic factors have become more important in
accounting for the location of industries today.   10 marks

(18 marks)

Answer (1) denotes correct point

(a) *In South Wales a number of physical factors led to the growth of early industry found there (iron and steel industry). These included the availability of raw materials (1), coal, iron ore and limestone (1) which were all easily accessible and found close together (1) in the Rhondda Valley (1).*

Another physical factor was location, they were close to ports (1), Swansea and Cardiff for easy distribution (1). The easy market for iron and steel to India through the British colonies also contributed to the growth of the iron and steel industry.

Comments

This answer refers to a named area and there is sufficient indication that it is based on a knowledge of the area. Reference to factors such as raw materials, coal, iron ore and limestone, easily accessible, in the Rhondda valley all obtained marks. Further references to the proximity of the ports at Swansea and Cardiff obtained additional marks.

References to the easy market in India, etc., was irrelevant and did not obtain any further credit.

The answer obtained a total of **6 marks out of 8**.

Answer

(b) *Human and economic factors have become more important for the location of industries today. The industrial estate at Bridgend (1) in South Wales is located there because it has a ready workforce (1), many of whom were unemployed due to the decline of the iron and steel industry (1). Businesses are attracted here by Government and EU grants (1) which offer things like a year's free rent or funding for machinery (1). The industrial estate at Bridgend also has good communication links with most of Britain as it is located near the motorway (1). Organisations like Marks and Spencer and Sony TV have factories here (1) as it has good links with the rest of the market and provides easy distribution and good market potential (1).*

Comments

The answer makes good reference to appropriate human and economic factors including the industrial estate at Bridgend with a ready workforce, from the declining iron and steel industry, the grants offered by government and EU, with examples, i.e. free rent and machinery funding, the closeness to motorways and finally the attraction of major companies such as Sony TV.

There is some repetition in the answer towards the last section which refers several times to the communication links and therefore no further marks are awarded for these comments. The answer has sufficient appropriate points to obtain a total of **8 marks out of 10**.

Parts **(a)** and **(b)** together therefore obtain a combined total of **14 marks out of 18**.

Topic glossary

Business parks: Areas which have become industrial estates with businesses which may sell products or provide services directly to the public.

Dereliction and decay: This refers to abandoned buildings, perhaps mines, offices or industries, which have resulted from closures. They often become a source of visual pollution if they are not demolished.

Diversification: This is the process whereby an economic enterprise such as a farm takes on a range of additional activities to increase profits.

Enterprise zone: An area which receives government assistance to attract new industry and create new employment opportunity.

Extractive industry: Primary industry which takes raw materials from the ground, e.g. mining, quarrying, fishing, forestry.

Greenfield site: Land which has not been previously used for industry or any other buildings.

Heavy industry: Industry which produces heavy bulk materials, e.g. iron and steel, textiles, shipbuilding.

High-tech industry: Industry which uses the latest technology to produce goods and services.

Industrial decline: This happens when for various reasons industries in an area have to close. This has mainly involved large traditional, heavy industry such as mining, iron and steel making, textiles and shipbuilding.

Industrial estate: An area set aside for modern, light industrial units.

Industrial growth: This happens when an industry increases its output, number of workers and profits and usually has a positive effect on the local community.

Industrial inertia: This occurs when an industry remains in an area long after the original location factors no longer apply.

Light industry: Industries which manufacture small, light bulk products, e.g. window frames.

Manufacturing industry: Industries which make a variety of products either finished or semi-finished.

Multiplier effect: This is the wider effect of change on other activities such as other industries, settlements or rural activities. For example, when a major industry closes this may cause local shops and other businesses to close.

Primary industry: Industries which are based on extracting raw materials such as coal, ores, farm produce and forests.

Science parks: These are industrial areas which are closely connected to technological institutions and universities and are often involved in producing high technology products.

Service industry: Also referred to as 'tertiary' industries which provide services such as retailing, wholesaling, transport, legal, administrative and trades for communities.

Sunrise industries: Term used to describe new modern highly technological industries such as electronics.

Urban Geography

Urban systems

You can be asked a question on the factors which influenced the choice of site of a particular settlement. You should be able to refer to a city in the UK which you have studied.

The **'site'** of a settlement is the actual land on which the original settlement was built.

Figure 8.1 illustrates a variety of settlement sites.

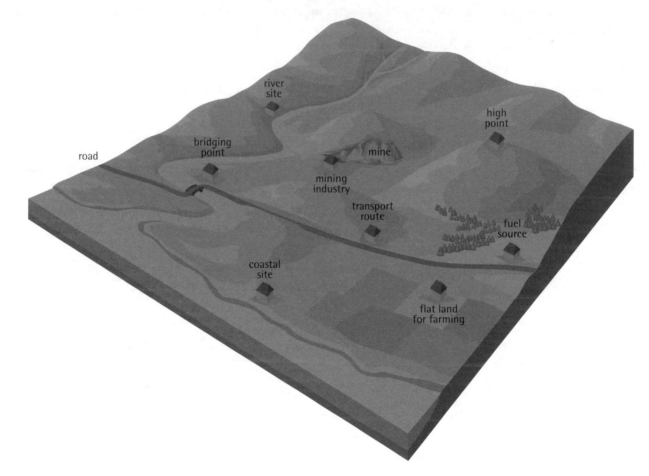

FIGURE 8.1: FACTORS INFLUENCING A SETTLEMENT'S LOCATION

Influential site factors include:

- A site which is near to a **supply of fresh water**, e.g. a river.
- A site which provides **flat land** for building on.
- Areas which have good fertile soils will be attractive for settlement.
- In earlier times a site which was close to a **source of fuel** such as a forest.
- Where the area was mainly marshy any area of **dry firm land** would be a suitable site. These are called **'dry point sites'**.
- A site which was **easily defended**, e.g. a high point, was often chosen.
- A site which lay **in a gap between upland and lowland areas** would be suitable. These sites are known as **'gap'** sites.

- Sites on the lowest point at which a river could be crossed were often selected. These sites are called **'bridging points'**.
- As industry developed sites close to **raw materials**, e.g. coal or iron ore, were popular.
- A site on a **coastal location** would offer opportunities for port and harbour developments.

The choice of site could be influenced by one or a combination of the above factors.

Settlements are also affected by their **situation** in relation to other places.

- For example, settlements on major routes often prospered through increased trade and customers.
- Settlements in areas where major routes converged often became major service centres, e.g. Stirling and Carlisle.
- Settlements in coastal locations often developed into major ports, e.g. Portsmouth.

Using map evidence to identify site factors

When using an Ordnance Survey map to identify site factors refer to the following evidence:

- **Contour lines (or lack of them)** will indicate the flatness or steepness of the land.
- **Nearness to a water source** such as a river is easily spotted.
- If the site is built on the meander of a river or on **high ground**, this should indicate **defence** as an important factor.
- Mines or quarries (sometimes currently disused) will reveal a local source of raw materials.

Case study

You should be able to describe the main site factors responsible for the growth of a city you have studied.

Top Tip

You must give the name of the city and refer to specific local features.

For **London** points might include **factors** such as

- the bridging point on the River Thames
- flat land to build on
- highly accessible by land and river for transport
- good water supply.

When discussing **London's growth** mention the following:

- Growth was linked to London's development as a **route centre**.
- Port facilties were built from Roman times and were developed through its history, enabling London to develop as a **major trading centre and port**.
- Good **road and rail links** to other parts of the country helped attract many people to the city.
- Being the home of government attracted many other institutions such as major banks, headquarters of major businesses and main law courts.
- Gradually London grew to become Greater London with a population of approximately 12 million people.

Settlement functions and land use zones

The word '**function**' refers to the various activities found in a city. These activities will be located in different areas of the city. These areas are called **functional zones**.

Quick Test

Make a list of site factors and write a sentence to show the influence of each factor.

The following types of functions and functional zones are found in most cities:

Residential

This includes all types of housing, including low, medium and high cost residential zones.

Industrial

This includes all types of industry found in particular zones in a city or town. Older industrial zones are found near the city centre while new estates can be located on former industrial sites or on the edge of the settlement.

Service

All cities provide services including retail and wholesale, trades, personal, financial and education. The most important service functions are found in the city centre area, also known as the **Central Business District (CBD)**.

Administrative

Services ranging from local town hall to national government offices are usually located in the CBD.

Educational

Colleges and universities, known as tertiary education centres, are usually found only in the larger settlements such as large towns and cities.

Transport

The transport function consists of the provision of transport services. This includes road and rail services with major railway and bus stations found in or around the CBD.

Entertainment

This includes cinemas, theatres, public houses, clubs and restaurants which are usually located in the CBD.

Are these zones located in any kind of pattern?

They are all contained in areas known as **functional zones**.

The location of these zones is closely linked to the value of land in different parts of the city.

Figure 8.2 shows that land values are highest at the city centre (CBD) due to the area containing functions which can compete for the **highest land prices**.

- Land values begin to **decrease** further away from the centre.
- Prices rapidly **decrease** in the area known as the twilight zone and area of low cost housing – mainly tenements.
- Further away land values begin to **rise** and are reflected in the better quality housing in the suburbs.
- At the **edge** of the city land values **decrease** once again and land is used for industrial estates, out of town shopping centres, council house estates.
- You should refer to these points and use examples of zones from a named city.

How do you identify the CBD on an OS map?

There are certain features you should look for on the map.

- As most maps show, the main roads converge on this area. You would refer to this and identify roads **by name** if possible.
- You would also look for buildings, such as train and bus stations, town halls, museums, information centres; perhaps colleges and other large further education establishments,
- However, the street patterns may show small narrow streets at the very centre indicating the oldest part of the settlement. Grid references of their location should be given to support identification.
- There may be some streets which have different shapes such as small cul-de-sacs and curves. Examples of the latter are found in Edinburgh and Glasgow.

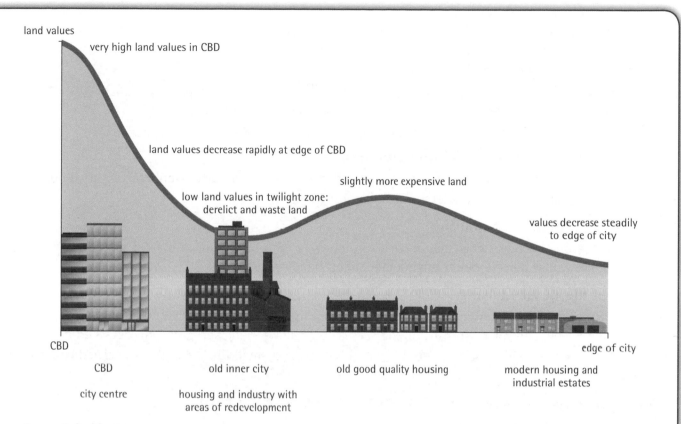

land values

very high land values in CBD

land values decrease rapidly at edge of CBD

slightly more expensive land

low land values in twilight zone:
derelict and waste land

values decrease steadily
to edge of city

CBD

edge of city

CBD

old inner city

old good quality housing

modern housing and
industrial estates

city centre

housing and industry with
areas of redevelopment

FIGURE 8.2: MODEL OF LAND VALUES AND URBAN ZONES ACROSS A CITY IN THE DEVELOPED WORLD

If asked to compare the CBDs of two cities, refer to:

- **Location** within each city.
- **General accessibility** by noting roads, railways, bus and train stations.
- The general size of the areas.
- **Types of zones** close by, e.g. industry or housing.
- The presence of **specific services** such as tourist information centres, churches, etc.

What are the main features of land use zones?

1 The core area – Central Business District

- The site of this area is usually the centre of the settlement and is the most accessible point.
- It is the area where most important services would locate to have maximum access to their customers.
- Functions such as retailing, wholesaling, offices, services, public administration, entertainment, transport, terminals, art centres are all located here.
- This would also be the area where land prices are most expensive.
- Few people live here, therefore few houses.
- There may be pedestrianised precincts.

2 The industrial zone

- Normally the site of industrial areas have certain common features such as flat land, accessibility in terms of roads, rivers, canals and possibly motorways.
- The older industrial areas are normally located close to the CBD.
- This area should contain a range of manufacturing industry including, for example, textiles, engineering, food-processing.
- This will contain large blocks of buildings often with the name 'works' beside them.
- If there is a river, there will be various industrial units along the banks including warehouses, docks, perhaps shipbuilding, oil refineries and power stations.

- Newer industrial areas may be seen either on the outskirts in the area known as the 'rural-urban fringe' in industrial estates, or near the city centre in redeveloped sites.
- Land values in new industrial areas are lower.
- Industries will be modern and contained in newer buildings with wider streets, perhaps lined with trees.
- The buildings will be smaller units than the large factories of the older zone.
- These industries will be based on perhaps electronics and small-scale manufacturing.

3 Area of low cost housing – inner city areas

- This zone should contain mainly less expensive, often nineteenth or early twentieth century, houses.
- These were built to house the workers of the industries in zone 2.

4 Area of medium cost housing

- As some people became richer they moved further away from the centre into areas where the housing in this zone would be of a better environmenal and building quality.

5 Area of high cost housing

- As more could afford to move even further away from the city centre a further zone of housing emerged, consisting of low density, high cost housing such as bungalows, detached and semi-detached houses.
- These were located either on the outskirts of cities or built in small rural villages now called **commuter villages**.

Why are the main features in some zones so different?

In answering this mention:

Housing

Street patterns help identify different residential zones.

- **Low cost houses** were built in a grid iron pattern often with small narrow streets.
- The environments of older low cost housing areas would not be very pleasant.
- The housing will almost certainly consist of tenements, which are high density housing.
- Many of these tenements may be showing signs of age and the effects of years of pollution from nearby factories.
- Many people have also been re-housed in schemes built on the boundary of cities.
- **Medium and higher cost** housing may be recognised from street patterns of curvilinear streets, cul-de-sacs and gardens.
- They will be areas of low density, wider streets, sometimes tree lined.
- Houses may consist of semi-detached and detached buildings.
- The overall environment will be much more pleasant with less traffic and less pollution than the inner city areas.
- There may be a wide range of services such as schools, small shopping centres, recreational areas and open space.

You may be asked to discuss in detail the main features of the CBD of a city you have studied. You could refer to the CBD of Glasgow. Its main features include:

- The Central Business District of Glasgow (Zone 1) is located on the northern edge of the River Clyde stretching a distance of approximately 4 kilometres west to east and 2·5 kilometres north from the river.
- This area contains all the services you would expect to find in any Central Business District, including large department stores, e.g. Frasers, Marks and Spencer, train and bus and subway stations and several large undercover shopping centres.
- Unlike some cities it also contains features such as public parks.
- The layout of the CBD is a grid iron pattern based around the area known as the Merchant City in the eastern edge of the CBD.
- The River Clyde is a barrier to further development to the south.

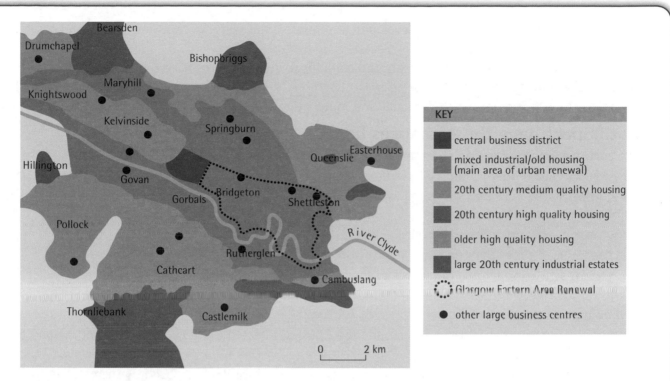

FIGURE 8.3: ZONES IN GLASGOW

Have any changes taken place recently in Glasgow's CBD?

Your answer should mention:

- As the population declined in Glasgow, many shops were affected.
- Many were no longer able to remain in business. Closure and for sale signs have signalled the closure of many famous stores.
- Many department stores have been replaced by the 'franchise' type of operation, e.g. Debenhams.
- Other changes included the building of undercover shopping arcades such as the Princess Arcade, the St Enoch's centre and Buchanan Galleries.
- Pedestrian zones are now common in areas such as Argyll, Buchanan and Sauchiehall Streets.

What were the reasons for change?

- Measures such as these have have helped Glasgow CBD to compete against out of town shopping and business complexes.
- Many centres such as the Braehead centre between Glasgow and Paisley provide a typical example of this type of new centre.
- Efforts to make Glasgow's CBD more accessible.
- Efforts to reduce road congestion, maximise building space and attract more custom to the city centres have included road improvements, e.g. one-way systems, and the renovation of old housing into offices.
- Other changes in the city's CBD have included renovation of old buildings such as the GPO into high cost flats to attract people back into the city.

What changes have affected the inner city area of Glasgow?

- Many new manufacturing units are replacing older industries such as heavy engineering works and textile factories in the old industrial zones.
- There have been changes along the banks of the river from the CBD to the western edge of the city which contain the remnants of Glasgow's **port industries** including warehouses, food processing industries, docks and harbours and shipyards.

- There has been a great deal of new development in these areas with new housing, hotels and exhibition centres such as the SECC being built.
- In the low cost housing areas, cul-de-sacs and curvilinear street patterns have replaced much of the rigid blocks of the grid iron pattern.
- The inner city environment is much better and much more pleasant than it had been previously.

What impact have these changes had?

- The environments of these areas have been transformed thoroughly.
- New estates with modern industry have been built.
- These new estates, however, do not always employ large numbers and unemployment often remains a problem.
- Many people now work from home using the advantage of technology.
- This has reduced time spent on travelling to work.
- This has made a considerable difference to their choice of residence.
- The increase in demand for housing in the rural-urban fringe or greenbelt area has seriously affected the environment of these areas.
- Changes to transport networks in and around the CBD has affected Glasgow ring roads and helped to reduce traffic congestion.
- Glasgow introduced traffic management schemes to reduce the problems caused by increases in traffic throughout the city. These included park and ride schemes outwith the city centre, one-way systems, bus lanes, parking restrictions and areas and buildings used for parking. These efforts helped to speed the flow of traffic particularly in the city centre.
- Traffic congestion remains a major problem in Glasgow especially during the rush hours.
- Some businesses on major routes have suffered through lack of customers because drivers can no longer park on main roads.
- Rengeneration schemes in some inner city housing areas have radically changed the quality of housing and local environment of many housing zones.
- Older properties have been demolished and replaced with new housing.
- Many families and older communities have been broken up and redistributed to areas throughout the city.

Top Tip

The case studies overlap with those in the applications section on Urban Management and Change. When revising refer back to these.

Quick Test

From memory list four changes in Glasgow's CBD and reasons for them and then do the same for the inner city area.

Geographical methods and techniques

Most techniques involved with urban geography attempt to identify and explain patterns of land use and change in urban areas.

Some of the analysis **uses Ordnance Survey maps** and will involve identifying some of the patterns discussed earlier in this chapter.

- Street patterns, the size and shape of buildings and identifying certain types of buildings from the key provided allow you to identify different zones on an urban map.
- Use these features also to compare areas such as different housing zones, the CBD and industrial zones.
- Your explanation of these patterns can be based on your knowledge of functional zones.
- Refer to transport patterns, road, rail and water to explain the location of service areas and industrial areas.
- Use your knowledge of urban models and urban zones to explain the pattern and the land uses within certain zones.
- Settlements regardless of their size attempt to serve customers within a specific area. The size and shape of that area is largely determined by the number of services provided.
- These service areas are known as the **settlement's sphere of influence**.

Sample question, answer and comments

Question
Study Reference Diagram Q1 (Figure 8.2 on page 79).

Referring to a named city in the Developed World which you have studied,

(a) describe and account for the likely land uses to be observed in Zone A, and

10 marks

(b) suggest why Zone B may have experienced considerable change in the last 30–40 years.

8 marks

(18 marks)

Answer (1) denotes correct point
(a) *Likely land uses to be found in the CBD of Glasgow are mainly service industries (1) e.g. shops, entertainment like cinemas, concert halls (1). Large offices are found here because of the easy accessibility (1). Many roads and railway lines meet in the CBD (1). Large department stores with big names are found here e.g. Marks and Spencers, Debenhams (1). Bid rents are very high in the CBD (1) because land is very desirable (1). There are many high rise buildings (1) because they do not take up space on the ground. There is not much space, no parks (1), it is a built-up area.*

Comments
This is a good answer which contains correct descriptions of land uses, within a named city, with plenty of examples of the types of land uses mentioned. Although several examples are given, only **1 mark** is given for each list. The answer also contains explanatory points as to why certain functions are there, e.g. 'accessibility' and 'bid rents' and 'lack of space'. These comments gave three relevant points in explanation. Overall the answer was good enough to obtain **9 marks out of 10**

Answer

(b) *Zone B has changed because of industry. Industry declined and new more modern industry moved out to new industrial estates on the outskirts of the city (1) e.g. Hillington (1). New housing schemes were also built on the outskirts (1) e.g. Castlemilk and Easterhouse (1) so people moved away. The invention of New Towns also got people's attention. There was housing and employment there (1), e.g. Irvine, East Kilbride, Glenrothes (1). In zone B, tenements were found. These have been rehabilitated by the GEAR project (1) (Glasgow Eastern Area Renewal). The old tenements were refurbished inside and out and 2 houses have been knocked into 1 to enlarge living space (1)'.*

Comments

The answer obtained marks for references to the changes to Zone b, including industrial change, movement of population to outer areas and the comments on the rehabilitation projects with named examples.

No marks were obtained by naming the outer city housing schemes. The comments on the New Towns were irrelevant to the answer since they did not refer directly to the changes within Zone B, and therefore did not obtain any further marks. The final comments on the refurbishment of older housing was worth an additional mark.

In total the answer obtained **8 marks out of 8**.

Overall parts (a) and (b) gave a total of **17 out of a possible 18 marks**.

Topic glossary

Central Business District (CBD): This is the zone which contains the major shops, businesses, offices, restaurants, clubs and other entertainments and is normally located at the centre of the settlement at the junction of the main roads.

Commuters: Those who live in commuter settlements.

Commuter settlement: Small settlements on the outskirts of major towns and cities where people live but travel into the main settlement for employment and services.

Dereliction and decay: Old buildings such as factories and houses which, through age, wear and tear, are no longer usable and have been abandoned.

Functional zones: These are areas of a settlement where certain functions are dominant, for example industrial zones.

Functions: These are individual activities which settlements perform such as commercial, industrial, administrative, transport, religious, medical, recreational and residential.

Green belts: Areas surrounding cities and towns in which laws control development such as housing and industry in order to protect the countryside.

High order functions: These are the most important functions such as department stores, council offices or art galleries.

Inner city: This is the area near the centre which contains basically the CBD, the older manufacturing zone and zone of low cost housing.

Low order functions: These are less important functions such as small corner shops, post offices and petrol stations.

Overspill population: This refers to people who have moved out of the main city to other smaller towns or new towns.

Pedestrianised zones: These are traffic-free areas within the city centre where people can shop and walk along streets where traffic is forbidden to enter.

Renewal and regeneration: This refers to the processes by which older areas are demolished and replaced by new buildings often having totally different functions to the original building or area.

Ring roads: Roads built specifically to take traffic away from the city centre and to help solve the problem of congestion.

Site: This is the actual land upon which a settlement was originally built.

Site factors: These are the factors which influenced people to choose a particular site such as nearness to water supply (river), flat land for building, high land for defence, at a suitable point on a river where a bridge could be built, near raw materials.

Sphere of influence: This is the area from which a settlement draws its customers for various functions. Usually the size of the sphere of influence varies in direct response to the size of the settlement.

Suburbs: Housing zones on the outskirts of towns away from the busy central inner city.

Traffic congestion: This is a heavy build up of traffic along major routes and within the city centre which causes great problems of cost and pollution for many cities in the UK.

Urban model: This is an idealised view of the internal structure of a settlement and can be used as a basis for comparing settlements.

Rural Land Resources

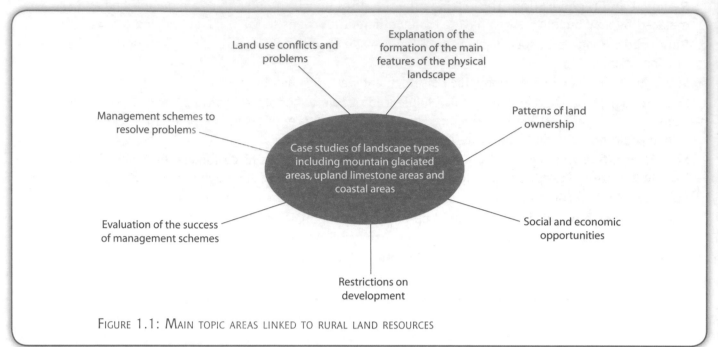

FIGURE 1.1: MAIN TOPIC AREAS LINKED TO RURAL LAND RESOURCES

Glaciated upland areas

For any **named glaciated mountain** area in the UK you should be able to describe and explain the formation of a number of landscape features.

It is important to know that in this section more marks are awarded in questions relating to landscape formation than in questions on the Lithosphere in Paper 1. For example there may be up to 20 marks available as opposed to 10 or 12 in Paper 1.

The features which you should know include features of **erosion** namely, U-shaped valleys, hanging valleys, pyramidal peaks, corries, corrie lakes (tarns), arêtes, truncated spurs, ribbon lakes, misfit streams, alluvial fans.

You should also know features of glacial **deposition**, which include moraines, eskers, drumlins, crag and tail, outwash plains.

Refer back to the chapter on the Lithosphere for a detailed discussion of these features.

What factors limit human activity in glaciated uplands?

- Due to the difficulty of the terrain and the poor quality of soils farming is limited mainly to hill sheep farming.
- Settlement and communication are restricted to valley areas where the land is flatter.
- Apart from quarrying a lack of raw materials has limited industrial development.
- Tourists are attracted to the splendid scenery of areas such as the Lake District and the Cairngorms for activities such as hill walking, mountaineering and skiing, despite their remoteness and inaccessibility.

Upland limestone areas

For any **named upland limestone** area in the UK you should be able to describe and explain the formation of a variety of landscape (surface and underground) features using annotated diagrams.

Features which you should recognise, describe and explain include surface features including intermittent drainage, limestone pavements, clints and grykes, potholes, scarp slopes and scree, and underground features including caverns, stalagmites and stalactites, and underground lakes.

Full descriptions and explanation of how they were formed are contained in the Lithosphere chapter.

Coastal areas

For a **named coastal area** you should be able to describe and explain the formation of features such as headlands, bays, stacks, caves, arches, spits, tombolos and longshore drift.

A full description and explanation of these features is also in the Lithosphere chapter.

For an upland area or coastal area you have studied, describe the opportunities which exist for social and economic development.

When answering this question you could refer to any of the national parks in England and Wales or those in Scotland since these provide good case studies for discussing social and economic opportunity.

Opportunities for development include property, industrial, recreational, tourism and commercial development.

Development will be restricted to those which meet the standards of national park legislation.

Developments such as industry or housing must not detract from the landscape in terms of, for example, visual pollution, e.g. quarries scarring the landscape.

These developments bring advantages including:

- Employment opportunity in, e.g. hotels, shops, building trades and the tourist industry in general.
- Tourists putting more money into the local economy.
- Improvements in the local infrastructure, e.g. road and railway.
- Improvement in local facilities, e.g. new libraries, clinics, leisure facilities.

Disadvantages of new developments include:

- Increased demand for new property, particularly holiday accommodation.
- Purchase of local housing for second homes.
- Increased traffic congestion and danger on local roads due to increase in visitors.
- Increases in various forms of pollution through litter, traffic, vandalism.
- Possible increases in land erosion, e.g. footpaths.
- Danger to farmland and forested areas through disregard for safety of livestock (gates left open, animals frightened by dogs) and fires.
- Pollution of water areas from water sport activities.

How can these problems be solved?

Your answer should mention:

- Conflicts avoided by national park authorities and other conservation agencies strictly enforcing **rules and legislation** on anyone causing deliberate damage.
- **Limitations** imposed on certain activities, e.g. water sports.

Environmental Interactions

FIGURE 1.2: NATIONAL PARKS AND MAIN CENTRES OF POPULATION

- Setting up **education centres** for visitors on how to respect the natural environment.
- The need to monitor activities continually to maintain the sustainability of national parks, e.g. employing park rangers.

Why can land ownership patterns make the management of National Parks difficult?

There is a wide variety of landowners within national parks, including farmers, golf courses, local industry (e.g. quarries), national government, householders, water board and owners of sites of historical and cultural importance, Ministry of Defence, Forestry Commission and National Trust. Many of these land users are in conflict with each other and with the aims of the national park authorities.

Case study: Loch Lomond National Park

Loch Lomond National Park was designated in 1997. The area was granted national park status in 2000 and Scotland's second national park, the Cairngorms, was designated in 2003.

Why was Loch Lomond given national park status?

Reasons included:

- The need to protect the physical environment of the area including the actual loch.
- The constant demand for development from a wide variety of land uses and activities, including water board, tourism, recreation, marina developments, communications, housing, industry and agriculture.
- The need to provide good access and facilities for public open air enjoyment and maintain established farmland.

Did this create any problems for the area?

- It allowed increased development of housing, recreational facilities such as camping and caravanning sites, and the use of the loch for increased recreational activities.
- Access through additional roads was increased.
- Agricultural land suffers from increased pressure from other land users.
- The physical environment is being spoiled by the increased number of visitors.
- Natural habitats of local wildlife is disturbed or destroyed in various ways.
- Water in the loch suffers increased pollution.
- Physical erosion of the landscape occurs as the number of hill walkers has increased.
- You can use these arguments in relation to any national park area, e.g. Lake District, Peak District or Yorkshire Dales.

Note that questions on this topic can refer to any upland area and are not confined to National Parks.

Case Study: Dorset/New Forest Coastline (Coastal area)

You should be familiar with land use conflicts in a named coastal area and the efforts made by various agencies to protect the area and be able to comment on the success of these efforts.

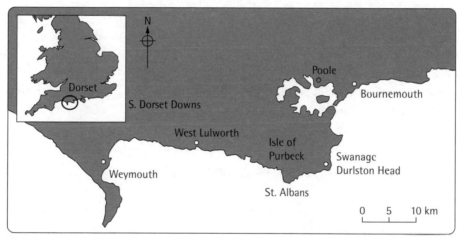

FIGURE 1.3: DORSET COASTAL AREA

Background

- In the New Forest Area in west Hampshire, the New Forest extends to the shore between the mouths of the Lymington and Beaulieu Rivers.
- There is a wide variety of land use present, including farming, industry, forestry, tourism, recreation and leisure.
- The estuaries of these rivers are centres for sailing, boat building and repairs.
- In the area of Southampton Water, much of the area is developed for major industry such as the Esso oil refinery and petrochemicals complex, housing developments and power stations.
- Coastal marshes in the area are used for nature conservation despite the risks from water pollution caused by industrial waste and domestic sewage.
- These activities have a major impact socially, economically and environmentally on this area of coastline.

Social, economic and environmental impact of the main land uses

Social impact is felt in terms of the area being a magnet for population since the area is very attractive to tourists, as a place to live, work and retire. 17 million tourists are attracted to the area annually and there are over 200 000 educational visitors catered for every year.

Economically the area benefits from the wide range of economic activities, including tourism, industry, forestry and farming. These activities generate a great deal of money for the people who live and work in the area.

Environmental effects of demands from tourism, port and ferry services and increased traffic congestion, include footpath erosion, demand for land for car parks and other amenities, threats to wildlife habitats, danger of marine pollution, conflicts between fishing, water sports and marine archaeology and the presence of the UK's longest onshore oilfield in the area.

Main land use conflicts within the Dorset coastal area

All of these activities combine to threaten the natural environment and to change the natural balance and ecological diversity of the area.

In addition the coastline is under threat from natural forces such as waves, currents and groundwater movement.

Management strategies to deal with issues

A document outlining the strategy for sustaining and improving the quality of the environment called the 'New Forest 2000' was published in 1990. Measures in this strategy include efforts to reduce pollution levels, protect the scenic beauty, improve the appearance of the coast, maintain the economy, protect the coastline, educate the public and to conserve features of historic and archaeological interest.

There have also been attempts to protect coastlines from flooding using dykes and flooding walls. The coastline needs to be managed to sustain human activities from the threat of marine erosion, to preserve coasts for conservation reasons and to preserve them from development. Responsibility for managing coasts generally lies with three agencies, namely the Environment Agency, MAFF (national government agency for coasts) and district councils.

Coastal defence strategies include measures such as seawalls, using large irregular rocks, gabions (wire baskets filled with rubble), groynes and embankments. Each of these measures has several advantages and disadvantages not least of which is cost. Some are relatively cheap while other such as seawalls can be very expensive to build and maintain.

Nature reserves have been created to protect wildlife and two country parks have been created to encourage sustainable tourism. Public authorities, which operate the nature reserves and the country parks, manage these strategies, and the area has achieved the status, if not quite the title, of a National Park.

Top Tip

When answering a question on this topic make sure that you mention a particular upland area or a named national park and identify specific places within them such as towns, villages and local landscapes, e.g. lakes or mountains.

These efforts in the New Forest Area have met with considerable success as have those in the other study areas discussed.

Role of public and voluntary bodies in the Dorset coastal area in dealing with land use issues

Public and voluntary bodies, other than those associated with nature reserves, involved in efforts to resolve land use issues include coastal management authorities.

- The area has been given Heritage Coast Status.
- It is protected by the work of the Environment Agency, National Trust, National Nature Reserves and has been designated as a Site of Scientific Interest and Special Marine Conservation Area.
- All of these bodies make great efforts to ensure that the natural environment of this area is protected as much as possible.

EU agricultural schemes

You might be asked to discuss the aims, advantages and disadvantages of any EU agricultural scheme you have studied.

In answering refer to:

- Set aside schemes whereby farmers are paid not to plant crops on up to 20% of their land.
- Industrial crops may be allowed but not golf courses.
- Aim is reduce surpluses in certain crops.
- The advantage is that farmers can access additional money for doing nothing with their land.
- The main disadvantage is that it imposes restrictions on types of crops produced, some of which might be profitable.
- Quota schemes which set a limit on the production of certain crops, e.g. dairy produce.
- Some areas are designated as Environmentally Sensitive Areas in order to protect them.
- Other areas may be declared Areas of Outstanding Natural Beauty and development of these areas may be highly restricted to preserve their natural beauty.

How effective have these schemes been?

- Overall effectiveness of schemes such as these varies throughout Europe.
- Some countries are more rigorous in enforcing the schemes than others.
- Food mountains and milk/wine lakes still continue to grow.
- It is difficult to achieve consistency due to different application of the rules throughout the EU.
- Those measures relating to restricting industrial, commercial and housing developments have had considerable success.

Sample question, answer and comments

Question

(a) For the Peak District National Park, or a named upland area you have studied: 8 marks

(i) describe the opportunities which this landscape provides for a variety of land uses; and   14 marks

(ii) explain the environmental problems and conflicts which may arise from the competing demands of these different land uses   (22 marks)

Answer (1) denotes correct point

(a) (i) *The Yorkshire dales national park boasts some spectacular scenery eg limestone pavement above Malham Cove (1). This attracts many tourists each year including school trips and people interested in geology (1). The villages in the national park are also very quaint (1) eg Malham (1) which also attracts tourists to the area. There are many recreational opportunities (1) eg potholing, hillwalking (1) mountain biking which also attract tourists.*

At Gaping Hill at some points in the year tourists are allowed to travel down on a chair lift to experience underground drainage (1).

With the area being made up of limestone, this attracts quarrying companies (1) eg Tarmac. As a result many quarries have been built in the area eg Swinden Quarry.

There is huge biodiversity of plant species in the hay meadows (1) and also lots of wildlife attracting nature lovers (1).

Environmental Interactions

There are hectares of grazing land available making farming one of the main land uses in the area (1). As a result of the hills, soils are thin and machinery cannot be used therefore hill sheep farming is practiced (1).

Comments

This is a very good answer which obtains **full** marks. Not only does it mention appropriate activities, it also states examples from the area, showing good knowledge of the case study. The answer obtains **8 marks out of 8**.

Answer

(a) (ii) *As a result of the quarrying, environmentalists believe that they are destroying a unique habitat (1) and since limestone is not a renewable resource it should not be quarried (1). Also many believe that Horton quarry (1) are not complying with the aims of the national park. Lorries leaving the quarry cause pollution (1) and dust covers plants at the side of the road (1) causing some species to die. (1)*

The environmentalists are in conflict with the tourists. Many paths to Malham Cove are heavily eroded (1) and paths to Ingleborough Fell have been eroded by mountain bikers (1). Tourists also leave litter which is dangerous for animals (1). Climbers on Malham Cove disturb falcons especially during the nesting season (1).

Environmentalists are in conflict with farmers since they believe they neglect barns and walls (1) and that they use too many chemicals and overgraze the land (1).

Comments

Again a good answer, covering the main conflicts with good examples being given. It shows good knowledge of the area and deserves a total of **12 marks out of 14**.

Topic Glossary

Conflict: This happens when two or more land users disagree as to the best use of the land. Often one land use is totally at odds with another such as industrial activities perhaps spoiling and polluting areas used for farming.

Conservation: This refers to efforts to maintain the basic beauty and attractiveness of areas both in the countryside and towns.

Country park: This is an area in the countryside surrounding a town or city which has been set aside for people to visit as a park.

Countryside: The majority of land area which has not been used for towns or cities.

Countryside Commission: This organisation has been set up by the government to monitor and protect countryside areas from harmful development.

Forestry Commission: This organisation is responsible for planting and looking after forests throughout the UK.

Habitats: Places where animals and birds live, e.g. hedgerows, fields and woodland. Often these are lost when land is developed.

HEP schemes: These are hydro-electric power schemes built for the purpose of creating electricity using water in reservoirs as a source of power to drive turbines. The electricity is fed into the national electricity grid.

Honey pots: Places which are highly attractive to tourists and are often very busy especially during the holiday seasons.

Land use: This refers to how humans make use of the physical landscape, e.g. forestry, farming, industry or settlement.

Land use conflict: This occurs when different activities compete with each other to make use of the land, e.g. farming and tourism.

National Park Authority: This is the organisation which looks after areas which have been set aside throughout Britain for the recreation and enjoyment of the public. Their aims also include protecting these areas of outstanding scenic beauty.

National Parks: Places of outstanding scenery set aside and protected for the purpose of attracting people from urban areas for leisure and recreational purposes. These areas are looked after by a government agency called the National Park Authority. There are eleven national parks in England and Wales and two in Scotland.

Rural: This is another name for countryside areas.

Settlement: Places where people live and work, ranging from small villages to large towns and cities.

Top Tip

Go back to the Lithosphere chapter's Topic Glossary for other landscape terms.

Rural Land Degradation

Figure 2.1 summarises the main topic areas linked to rural land degradation.

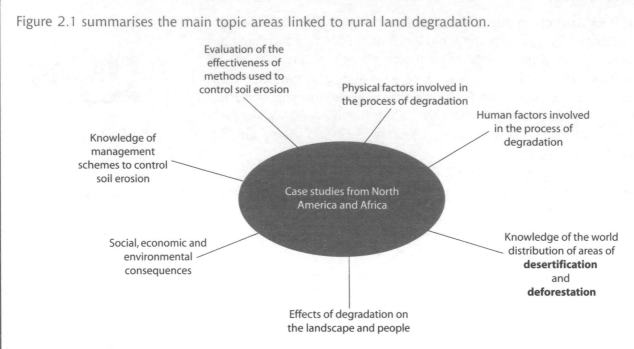

FIGURE 2.1: MAIN TOPIC AREAS LINKED TO RURAL LAND DEGRADATION

What factors are responsible for severe land degradation?

Physical factors include:

- Soil erosion
- Wind erosion
- Erosion by water
- Reduction of rainfall

Human factors include:

- Overcropping
- Overgrazing
- Monoculture
- Deforestation
- Mining
- Inappropriate farming techniques

In the examination you will be asked to refer to a case study from either North America, Africa north of the equator or the Amazon basin.

Figure 2.2 shows the processes involved in land degradation.

Case Study – USA

For the Dust Bowl in the USA during the 1930s:

Physical factors include:

- Rainfall was reduced because of distance from the sea.
- Rainfall was also reduced by a rain shadow effect which was created by the Rocky Mountains.
- Moist air from the Gulf of Mexico during the summer sometimes veered north east causing a loss of rainfall in the Great Plains.

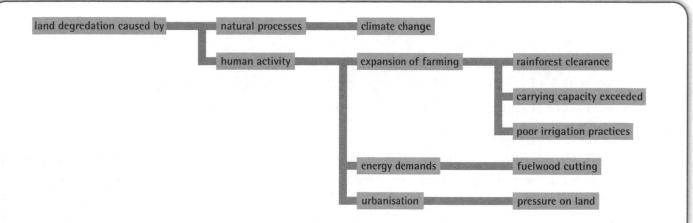

FIGURE 2.2: PROCESSES INVOLVED IN LAND DEGRADATION

Human factors include:

- **Monoculture** depleted the soil of moisture and nutrients.
- **Soils** were ploughed too deeply.
- **Ploughing** of marginal land during wet years left them fragile during dry years.
- Small farms overcropped during periods of low prices.
- Land was also overgrazed.

For the Tennessee Valley, USA:

Physical factors include:

- The presence of many steep slopes throughout the area.
- Rainfall from heavy storms rather than a more regular pattern.

Human factors include:

- **Deforestation** left the soils exposed to erosion.
- **Mining and farming** removed vegetation thus increasing soil erosion.
- Overcropping which weakened the soil further.
- Monoculture further damaged soils.

Make a list of three physical and three human causes of land degradation and remember these.

What are the main effects of wind erosion?

- Wind erosion blew away soil weakened by agriculture.
- Wind erosion stripped land of nutrients and reduced crop growth.
- In the 1930s millions of hectares of farmland were lost due to the worst example of wind erosion in the 'Dust Bowl'.

Describe rainfall patterns and their effects

When answering mention:

- Areas of high/low (deficit) rainfall.
- High and low seasons
- Long periods of low rainfall.
- Exposed soil being eroded by wind or flash floods.
- Remaining soils losing fertility and nutrients.

What are the main effects of land degradation on the local population?

These include:

- Farmers cannot graze animals.
- Soils are unable to maintain crop growth.
- Famine can occur due to low crop yields.

Environmental Interactions

- People migrate to other areas.
- The disappearance of traditional farming.
- Vegetation removal leads to global warming and the greenhouse effect.
- Vegetation removal increases ground water run-off rates often leading to flooding.

You will be asked to discuss the impact of land degradation in a specific area you have studied.

Case Study – Amazon Basin

In the Amazon Basin land degradation caused:

- The destruction of their way of life for local people.
- The destruction of activities such as rubber tappers.
- Reduction of fallow periods which resulted in reduced yields leading to food shortages.
- Conflict between big business and locals.
- Migration of local people from traditional habitats.
- Increase in poverty and social deprivation.

Case Study – North Africa

Because of land degradation in Africa north of equator:

- Crop failures led to major famines, e.g. Ethiopia/Sudan.
- There are mass migrations often to refugee camps.
- Traditional farming, e.g. pastoral nomadism, collapsed due to overgrazing.
- Pressure on land increased as nomads settled in villages.
- Widespread poverty resulted in increased mortality rates, especially infant mortality rates.
- There is a breakdown of soil structure which caused the Sahara desert to increase around its edges.
- There is increased wind erosion of dried out soils.
- There was further erosion from rainfall.
- There is a lowering of water tables

Top Tip

Look at figure 2.3. Beginning with letter A write down one feature of soil conservation shown and if answering a question on this topic refer to this list. For example A: Afforestation C: Contour ploughing.

How do soil conservation measures help to reduce land degradation?

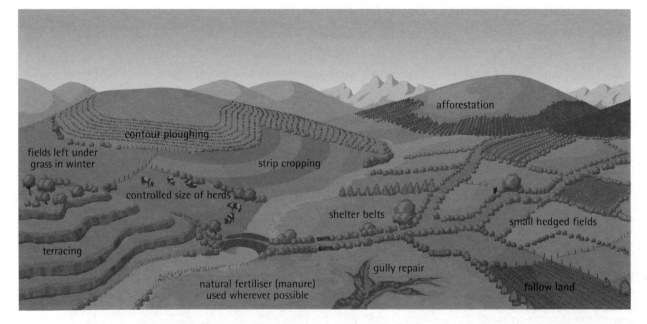

FIGURE 2.3: SOIL CONSERVATION MEASURES

These measures include:

- Building dams.
- Planting trees as windbreaks.
- Stabilising dunes with grass.
- Terracing slopes.
- Improving irrigation.
- Controlling grazing and fencing off areas to prevent vegetation removal.

In the Amazon Basin measures included:

- Reforestation with mixed trees.
- The use of crop rotation by farmers.
- The purchase of forest areas by conservation groups.
- Returning forests to native people.
- Attempts to control deforestation through government legislation

In North America measures included:

- Contour ploughing and crop rotation.
- Extensive use of fertilisers.
- Widespread access to irrigation schemes.
- Leaving fields fallow.
- Government subsidies for farmers
- Improved mechanisation.
- Government intervention, e.g. subsidies.
- Soil protection being given high priority.
- All of these measures have been very successful but very expensive.

Geographical methods and techniques

In some questions you may be asked to interpret graphs showing temperature change or rainfall variation over a number of years. When doing this, refer to:

- General patterns or trends.
- Specific years where change involved an increase and a decrease in the figures.
- Specific figures.
- Note differences between years by calculating specific change either plus or minus.
- If average figures are given, note periods when the change is significantly higher or lower than average.

Topic glossary

Cattle ranching: Rearing of large herds of cattle on areas of cleared forests to provide beef to be sold for export.

Deforestation: Removal of trees usually on a large scale.

Degradation: The process of reducing land which was formerly productive into unproductive land.

Desertification: Process of turning land which was formerly productive in desert.

Environmental Interactions

Drought: Prolonged periods without rainfall which may last from several months to several years.

Food chain: A system whereby various forms of life provide food for each other starting at one point and finishing at another, e.g. from small fish to larger and eventually to humans.

Global warming: The heating up of the earth's atmosphere by the sun's rays due to the effect of atmospheric pollution.

Greenhouse effect: A rise in temperature in the earth's atmosphere due to the effect of increased carbon dioxide and other gases in the atmosphere from various sources such as industrial pollution, burning forests, car exhaust fumes and smoke from domestic chimneys.

Insecticides: Chemicals used by farmers to kill insects which feed on crops.

Irrigation: An artificial way of providing water for farming from sources such as rivers, wells, canals, field sprays.

Logging: A commercial business which cuts down trees to provide timber for sale for different purposes.

Overcropping: Growing crops continually to the extent that the soil becomes exhausted of nutrients, becomes infertile and unable to sustain further growth.

Overgrazing: Allowing animals to overeat grass to the extent that the underlying soil is exposed and cannot sustain further growth.

Marginal land: Land which is on the outer limit of sustainable growth and development where crop growth is only just possible.

Nomadism: The system whereby people migrate with animals throughout the year to find new areas for grazing livestock also known as 'Pastoral Nomadism'.

Pollutants: Material which is released into the environment which ultimately causes damage to the physical landscape and atmosphere.

Reforestation: This is the process of replanting trees in former forested areas.

Shelter belt: A line of trees planted to provide shelter from the wind for fields by interrupting the flow of wind.

Soil conservation: Attempts to protect soil from damage using methods such as fertilisers, irrigation, shelter belts and ploughing along contours and using terraces to conserve soil.

Soil erosion: The process by which the top-soil is removed leaving the land infertile.

Windbreaks: Trees which interrupt the flow of wind so as to protect fields and their crops.

Top Tip

When answering a question on a topic such as this, always turn over to the next page of your exam paper to ensure that there is not a further part to the question overleaf.

River Basin Management

Figure 3.1 summarises the main topic areas linked to river basin management.

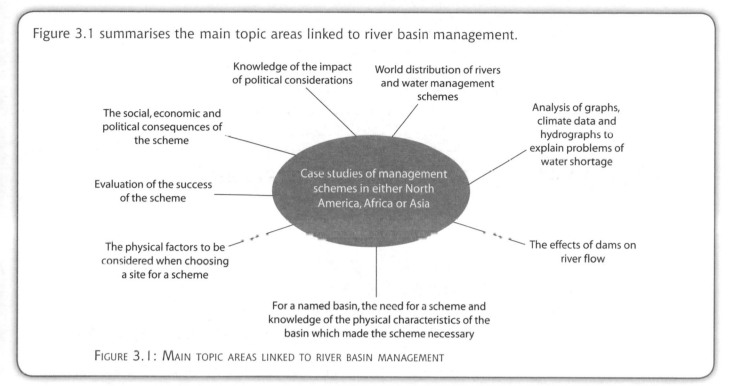

FIGURE 3.1: MAIN TOPIC AREAS LINKED TO RIVER BASIN MANAGEMENT

Distribution of river basins

In the examination you may be given a map and asked to describe and explain the distribution of river basins and water control projects with reference to either North America, Africa or Asia.

Figure 3.2 shows major river basins in Africa and North America.

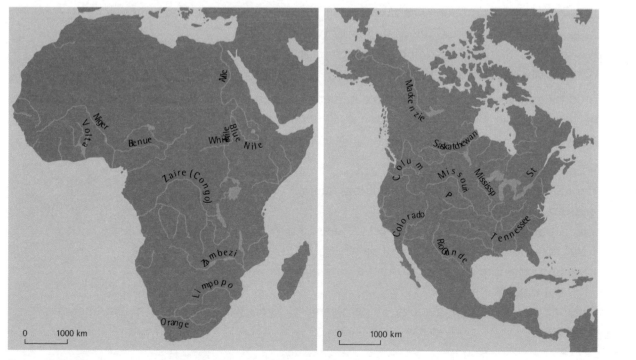

FIGURE 3.2: MAJOR RIVER BASINS IN AFRICA AND NORTH AMERICA

Environmental Interactions

In your answer, depending on the area chosen, refer to:

- General patterns of distribution and the number of rivers.
- Directions of flow.
- Explain the distribution by referring to mountain ranges as major sources of rivers due to their greater rainfall.
- Climate patterns throughout the area.
- Places where river basins meet sea/oceans.

Water supply and management

You should be able to explain, using suitable resources for any river basin in North America, Africa or Asia, why there is a water supply problem and why a water management scheme is essential

Figure 3.3 shows data for the Nile basin.

Referring to the figures given for the Nile Valley (Figure 3.3):

- Note the wide seasonal fluctuations in discharge levels.
- Refer to the steady flow of 100–500 million cumecs between November and May, rising to 200 million cumecs in June.
- Note a rapid rise in July to a peak discharge of 900 million cumecs in late August/early September, with an equally sharp decline in October.

When explaining this, refer to:

- A July–September surge due to pronounced seasonal regime of the Blue Nile's catchment area (see climate graph for Bahr Dar).
- The more regular flow of the White Nile helps compensate for the dry season in Ethiopia.
- Water levels are maintained at a steady level for the rest of the year.

What effect has the Aswan Dam had?

- The flow of the river has been even and regular since the Aswan dam was built.
- Extremes giving rise to annual floods no longer occur.
- Despite slight fluctuations, maximum discharge rarely exceeds 250 million cumecs.
- The dam clearly controls river flow.

You should be able to explain, referring to a named river basin, e.g. River Nile or Colorado, which you have studied, why there was a need for river basin management.

In your answer mention climate, landforms and environmental problems.

Depending on the basin chosen refer to:

- The rainfall pattern, especially periods of low rainfall.
- Any seasonal droughts.
- Any navigation problems due to physical features of river valley.
- The need for water transfer to areas of deficit.
- The irrigation potential and flood control during high rain season.
- Any HEP potential.
- The suitability of underlying geology (impermeable rock) for water storage schemes and dam construction.

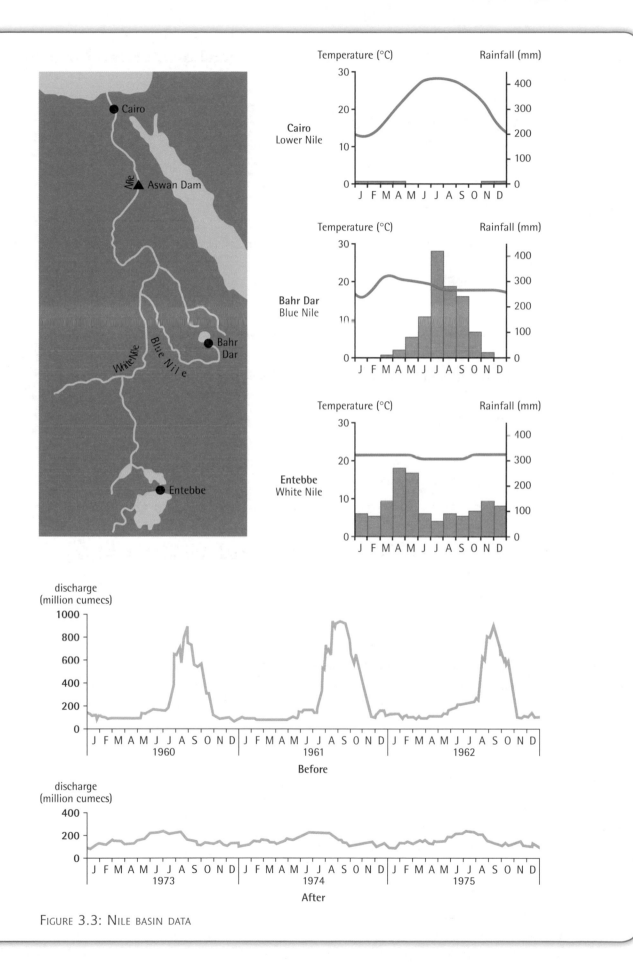

FIGURE 3.3: NILE BASIN DATA

Environmental Interactions

What physical and human factors influence the choice of site for dams and reservoirs?

In answering refer to:

- The need for solid foundations for a dam.
- The need for a narrow cross-section to reduce dam length.
- A large deep valley to flood behind the dam.
- A sufficient water flow.
- Local evaporation rates.
- The permeability of underlying rock.
- How much farmland would be flooded.
- How many settlements would be flooded.
- Local population distribution.
- The availability of HEP or irrigation.

How does the scheme affect the hydrological cycle?

In answers refer to:

- Increased evaporation from the surface area of reservoirs.
- Local climate.
- Less water flowing below dams.
- Diverting of rivers.
- Changing water table levels.
- Infiltration rates being affected by water in reservoirs.

What social, economic and environmental effects does the water control project have on the basin?

How successful is the scheme in terms of social, political and environmental impact?

In your answer mention:

Social benefits

- Population provided with increased food supply.
- Less disease and poor health due to better water supply.
- More recreational opportunities.
- More widespread availability of electricity.

Adverse social effects

- People will be forcibly removed from valleys.
- Water borne diseases could increase, e.g. schistisomiasis.

Economic benefits

- Improved yields in farming.
- Industrial development is helped by HEP schemes.
- More water for industry.
- Improved navigation channels.

Adverse economic effects

- The new scheme would be very expensive.
- Dependence on foreign money results in increased debt.

Environmental benefits
- Fresh water supply is increased.
- Sanitation and health are improved.
- Flood control.

Adverse environmental effects
- Water and industrial pollution increases.
- The silting of reservoirs increases.
- Salinity rates further downstream increase.
- Historical sites may be flooded.

What political problems are created by the project?

These include:

- Dependence on neighbours upstream.
- Water sharing by different states or counties requires complex legislation.
- Reduced flow and increased salinity.
- Problems over allocation of shared costs.
- Increased pollution across borders.

You could use the above points as a basis for answering questions about social, economic and environmental benefits and the adverse consequences of water management projects.

You must refer to specific examples from a case study which you have studied, e.g. a named dam or reservoir, local settlement and industry, types of farming in the area which have benefited from the project. This indicates that you have specific rather than general knowledge of the topic.

Topic glossary

Drainage: This is the term used to describe all surface water in a river system and should not be confused with underground pipes used to drain water from bog or marshland.

Environmental consequences: This refers to the net effects on the physical and human environments of changes to river basins through physical or human changes.

Infiltration: The process by which water from precipitation seeps into the soil and sub-soil.

River basin: This is the water catchment area of a river and includes the main river and its tributaries.

Salinity: This is the salt content in surface water such as rivers, streams and lakes.

Seasonal fluctuation: This happens when the normal climatic pattern is interrupted, for example when there is a change to rainfall distribution which could result in drought.

Silting: This is the amount of sand and other material carried in solution in rivers and, when deposited, can reduce river flow.

Tributary: This is a smaller river or stream which runs into a larger river.

Water control project: This is the name given to efforts to manage various aspects of a river basin such as river flow, reservoirs, silting, navigation and river discharge through, for example, the use of dams, canals, aqueducts and diversion schemes.

Urban Change and Management

Figure 4.1 illustrates the main topics related to case studies of cities in More Developed Countries and Less Developed Countries.

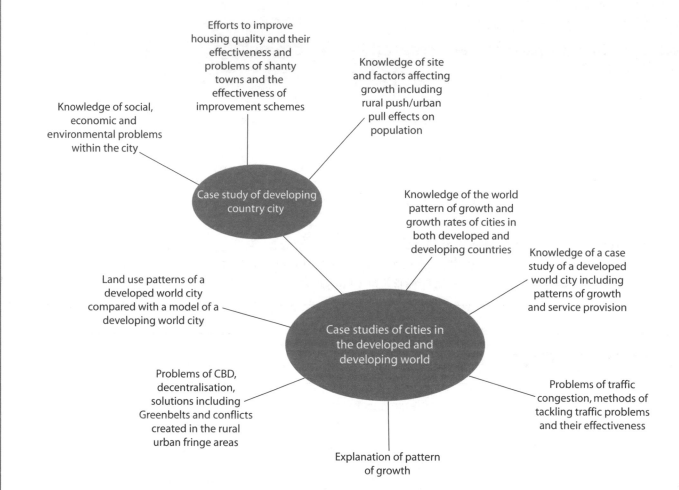

FIGURE 4.1: MAIN TOPIC AREAS LINKED TO CASE STUDIES OF CITIES IN MORE DEVELOPED COUNTRIES AND LESS DEVELOPED COUNTRIES

Exam questions often begin by asking you to look at either maps or data and to describe and explain the pattern of urban growth throughout the world. Figure 4.2 shows a typical diagram.

In answering this you should refer to specific cities and countries.

Figure 4.3 shows city distribution in the USA.

Prior to 1950, cities in **developed** countries had the fastest and largest growth, e.g. London, New York and Paris.

Reasons for this include:

- High birth rates, especially in cities.
- Decreasing death rates due better health care.
- Increasing employment opportunities in industry.
- Migration from rural to urban areas.
- Emigration from 'old world countries' in Europe to the cities in, e.g., United States, Canada and Australia.

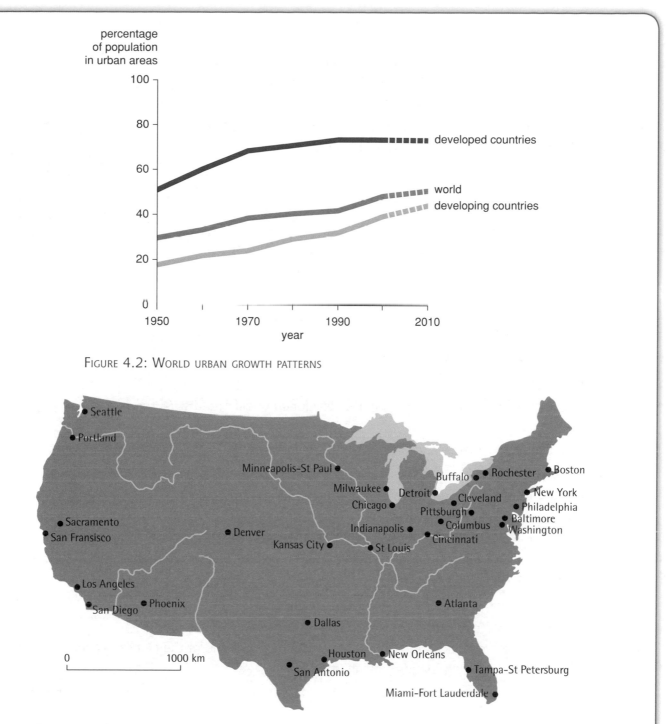

FIGURE 4.2: WORLD URBAN GROWTH PATTERNS

FIGURE 4.3: DISTRIBUTION OF CITIES IN THE USA

After 1950 the pattern changed. Growth was rapid and highest in cities in **developing** countries.

This happened because of:

- A population explosion in these countries due to increasing birth rates and decreasing death rates.
- Improved medical care provided by aid agencies.
- Increase in manufacturing industries in several countries.
- More employment opportunities within cities.
- Increased migration to cities from rural areas.

Environmental Interactions

Distribution of cities

You are often asked to select a country in the developed or developing world and describe and explain the distribution of cities as, for example, in Figure 4.3 – cities in the USA.

In this example city distribution is related to:

- Choosing coastal locations such as Florida and California.
- Choosing sites on natural routeways through valleys.
- Choosing sites close to raw materials.
- Choosing areas with good climates.
- Avoiding mountain and desert areas.

In Brazil city location was influenced by:

- Choosing suitable coastal sites, e.g. Rio de Janeiro and São Paulo.
- These sites had flat land and were highly accessible.
- The sites had advantages for developing port and fishing industries.
- The presence of the rainforest limited city development inland.
- Brasilia, which was purposely built inland to encourage the development of the forested areas, is an exception.

Top Tip

In the core Urban Geography chapter we discussed how to describe and explain variations in land use values within cities. Refer back to this.

Growth of cities

You can be asked to refer to a named city in the developed world and discuss the factors responsible for its growth.

For New York factors included:

Human factors

- The historical development with early settlers arriving in this area from European countries.
- Nodal position for trade with Europe.
- Development of major industries, e.g. textiles, port industries.
- Being an immigration point for people arriving from European countries and settling in the city during the nineteenth and twentieth centuries.

Physical factors

- A deep water location for large ships.
- The city having a commanding strategic position at the seaward end of the Hudson-Mohawk gap.
- The city having been built on islands – Manhattan.

Similarly you could be asked to describe and explain the factors involved in the site and growth of a city in the developing world.

In São Paulo, Brazil, site factors included:

- Choosing a site on a plateau near the confluence of the Tiete and Tamanduate rivers.
- Choosing a site 750 metres above sea level.

Quick Test

List the main reasons for the growth of cities in developed countries.

- Choosing a coastal location offering advantages for industry and port development.
- Population increases due to high birth rates and decreasing death rates.
- Migration from rural areas. This was due to factors such as low incomes, lack of education, employment and general low standards of living.
- Pull factors included industrial employment with improved income, better housing, education and health opportunities.

What problems did this city face as a result of growth?

Problems included:

- Expectation of incoming population being too high and people being disappointed.
- Homelessness due to lack of housing.
- Families either living on the streets, in squatter areas or in shanty towns.
- Lack of proper facilities such as clean water, electricity and sewage disposal schemes.
- A high incidence of health and disease problems.

What are shanty towns, where are they found and what are their associated problems and what efforts have been made to improve conditions within them?

Location and problems

- Shanty towns are usually found on the outskirts of cities.
- Known as **favelas** in São Paulo, these contain vast areas of poor, basic housing.
- The areas have emerged due to homeless immigrants and high birth rates causing rapid population increases combined with a lack of housing.
- Houses lack sanitation, clean water, electricity, cooking facilities and may be constructed of the most basic materials, e.g. wood.
- The shanties lack services, proper roads and are usually located along major routes into the city.
- People may have to walk long distances to get to the city.
- Some people may actually prefer to live in shanty towns as opposed to the insecurity of life on the streets.
- They provide contact with the extended family.

Efforts to resolve these problems include:

- City authorities have built satellite towns.
- Loans and grants are given to help build new houses or improve existing houses.
- There are improvements to local infrastructure, e.g. power, water, sewerage and roads.
- Authorities have built new housing estates with low cost housing and basic facilities.

Have these efforts been successful?

These efforts meet with difficulties such as:

- Increasing costs.
- The reluctance of inhabitants to move.
- A breakdown of community and family units.

Quick Test

Write a list of factors which can affect the choice of site of a city.

- High rent in new housing areas
- As improvements are introduced population increases in the shanty towns.
- There are still thousands of people living in shanty towns in cities in Africa, South America and Asia.

How have cities in the developed and developing world changed?

In economically more developed countries many changes have taken place in cities including population decreases, changes in housing, changes in industry, environmental changes and changes in service provision and shopping patterns.

Similar changes have occurred in cities in less developed countries especially in relation to population growth, infrastructure, industrial growth, water and sanitation provision, housing provision, education and health standards.

How have these changes been managed and what problems have arisen?

City authorities throughout the world have attempted to manage change through planning controls, building programmes, industrial renewal and the introduction of traffic management schemes. These strategies have all had varying degrees of success.

Change due to industrial decline

- Many industries have declined and the land they were built upon has been abandoned.
- Many of these pieces of land have been bought by property developers who have introduced development schemes including new housing and shopping areas.
- Land near the city centre offers attractive sites for development.
- Cities have large areas of dockland which is no longer in use.
- In Glasgow, Liverpool and London these docklands now contain a wide variety of new land uses such as new housing, hotels, shopping and tourist centres.
- The environments of these former industrial areas have improved immensely.
- Sites of former manufacturing industry have been developed as new industrial/trading estates and science parks.

Housing changes

- Many parts of cities have buildings which through age and lack of care have fallen into disrepair.
- Their visual appearance may be unsightly and more importantly they may be unsafe for use or habitation.
- Common sites for them include being near railway lines, inner city housing areas and along the banks of rivers.
- Many cities have made great efforts to remove these buildings and replace them with newer buildings.
- New buildings include office blocks, new housing, and undercover shopping centres.
- These changes have often been achieved through the financial assistance of government and European Union grants.

Environmental changes

- Pollution from industries discharging waste into rivers, urban blight caused by derelict buildings, air pollution from industries and traffic has affected many cities.
- Measures which have been taken to tackle these problems include the introduction of smokeless zones, building of high chimneys, banning of industries which are the source of obnoxious smells or pollutants.
- City councils employ environmental health departments to implement laws and regulations to monitor and control pollution.
- Many cities have improved the quality of the environment to the extent that they have won government awards.

What causes an increase in traffic congestion in cities?

This is due to:

- An increase in commuter traffic from rural-urban fringe areas.
- Poor quality roads within inner city area.
- People in developing world cities moving into inner city areas for employment.
- Inadequate public transport services.
- Low investment in infrastructure within developing countries.

What attempts have been made to reduce traffic congestion?

Measures taken include:

- **Changes to road systems**, including ring roads; one-way systems; use of 'bus only' lanes; contra-flow systems; widening of roads; building by-passes.
- **Parking restrictions** to prevent parked cars blocking main routes.
- Providing **alternative** means of transport such as park and ride schemes, cheap public transport and road and rail links.
- Additional **congestion charges** on cars.
- Alternatives such as out of town shopping centres.
- Authorities have to balance their solutions very carefully.

Top Tip

In the core Urban Geography chapter we discussed changes and problems affecting the CBD of Glasgow. Refer back to this topic.

Why have many people moved away from cities and what problems has this caused?

- The population of many cities has decreased as many people left to live in various settlements outside the main city.
- Large numbers of people were prepared to pay the cost of increased travel in terms of time and money to live in what they considered a better environment.
- The populations of these former rural villages have often grown enormously.
- In other cases, large numbers of people from cities were offered the chance to move to New Towns with the promise of employment as well as housing.
- These new towns were very popular.
- The result of these movements was that population of cities and the income from this population decreased alarmingly.
- Cities need people to support their services.
- Loss of population due to development of new towns and commuter belts have created financial problems for city councils.
- On the rural-urban fringe new housing and new industrial estates have led to urban sprawl and loss of rural land.
- This has caused conflict between new and traditional rural activities.

What attempts have been made to reverse or contain this movement?

- Many cities have tried to attract population back by investing in new and better housing, shopping areas and by trying to attract new industry.
- A great deal of money has been invested in upgrading older properties close to the city centres.
- The idea is to attract more and more people from the commuter settlements into the city.
- The process of re-populating these inner city areas has been termed **'gentrification'**.
- **Greenbelt legislation** introduced in the 1950s was an attempt to curb development in rural fringe areas and has been successful.
- Limitations on the number and type of buildings have protected rural environments.

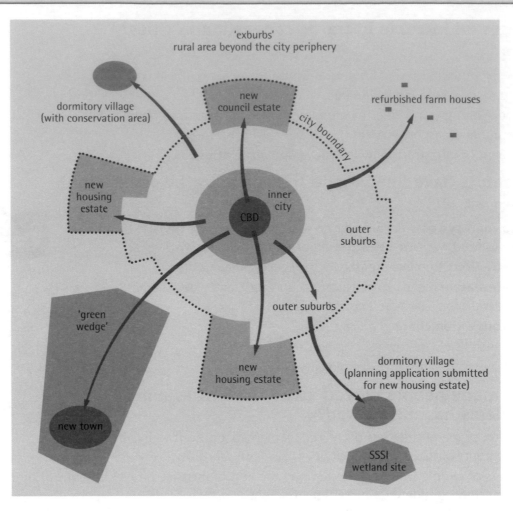

FIGURE 4.4: OUT-MIGRATION PATTERNS AROUND A DEVELOPED WORLD CITY

Land use patterns

You are often asked to compare and contrast land use patterns in a city in the developing world you have studied with that shown on a simplified map.

Figure 4.5 shows such a map.

What measures have been taken to improve the quality of life in developing world cities?

Measures include:

- Cities being involved with one of the international aid schemes which provide money, medical and educational services for local communities.
- These include United Nations sub agencies, e.g. United Nations Education Social and Cultural Organisation (UNESCO), World Health Organisation (WHO) and World Bank.
- Introducing primary health care schemes and local self-help schemes, particularly in the shanty towns.

Success rates vary from country to country, city to city within the developing world.

Top Tip

You must refer to specific locations within your chosen city to show detailed rather than general knowledge.

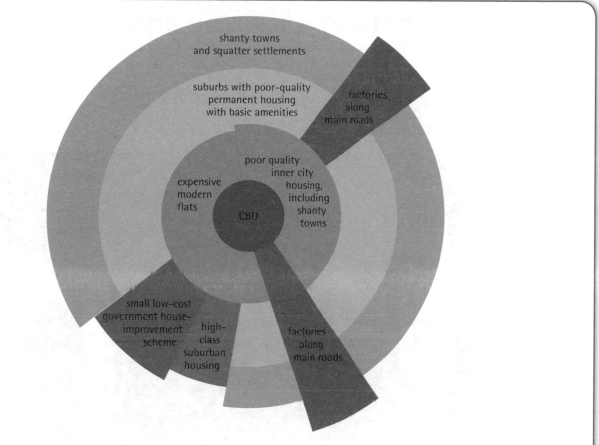

FIGURE 4.5: MODEL OF A DEVELOPING WORLD CITY

Why are there variations in housing quality and what has been done to improve poorer residential areas?

How effective have these efforts been?

Reasons:

- Housing quality varies due to lack of investment.
- Many houses lack basic amenities, particularly a supply of clean water, sewerage, electricity and basic cooking facilities.
- Conditions may be extremely basic and unhealthy.
- Some areas may be inhabited by squatters.
- This contrasts greatly with other richer parts of cities which have high quality housing, all necessary facilities, healthy and clean environments.

Improvement efforts

- In São Paulo efforts have been made to re-house people in new town areas.
- Improvements to water supplies in many residential areas have been treated as a priority.
- Self help schemes to improve local infrastructure using local labour have been backed by city authorities.
- Materials have been supplied to help improve the basic fabric of the poorest quality houses.
- In São Paulo financial incentives have been offered to residents to move to other parts of the city.
- Wells provide clean water.
- Sanitation has been provided, as has garbage disposal.
- These have helped to improve the quality of life and reduce levels of disease.

Sample question, answer and comments

Question

(a) With the aid of Reference Photograph Q4 and referring to a named city which you have studied in an ELDC (Economically Less Developed country):

(i) describe the social, economic and environmental problems created by shanty towns; and

(ii) describe ways in which such problems are being tackled.   18 marks

Answer (1) denotes correct point

(a) (i) *In Sao Paulo shanty towns have no sanitation or sewage removal schemes (1), raw sewage causes disease (1) such as typhoid and cholera (1) which spreads quickly through closely packed houses in favelas (1). The people get these diseases through unclean drinking water and vermin (1). The favelas are built on hillsides which could collapse after heavy rainfall (1). They are illegally built (1). Crime rates are high due to poverty (1), drug trafficking, prostitution etc. (1). There are no buses from these areas into centre for workers (1). Workers do low paid jobs such as shoe shiners (1). Edith Gardens is a favela in Sao Paulo (1).*

Comments

This is a very good answer showing good knowledge of the case study. The answer refers to the main features of the shanty town, problems and gives good examples, e.g. 'favelas' and 'Edith Gardens' and scores a total of **10 marks**.

Answer

(a) (ii) *Industry is being dispersed outside the city (1) and cheap housing is built there so some shanty town people move there and are employed in industry (1). Satellite towns are built with better facilities (electricity, clean water supply) (1). People in the shanty town are given houses and will often improve the area themselves (1). There will be a major clean up project of the River Tiete and Pinheros (1) so they can be used for water (1).*

Comments

This is another good answer which mentions the main points. The answer finishes with reference to named examples, which confirms good knowledge of the case study. In total the answer gains **6 marks** and taken together both answers achieve **16 marks out of 18**.

Topic glossary

Central Business District (CBD): This is the zone which contains the major shops, businesses, offices, restaurants, clubs and other entertainments and is normally located at the centre of the settlement at the junction of the main roads.

Commuter settlement: Small settlements on the outskirts of major towns and cities where people live but travel into the main settlement for employment and services.

Commuters: Those who live in commuter settlements.

Dereliction and decay: Old buildings such as factories and houses which through age, wear and tear are no longer useable and have been abandoned. They cause visual pollution and are often demolished to make way for new developments.

Functional zones: These are areas of a settlement where certain functions are dominant, for example industrial zones.

Functions: These are individual activities which settlements perform such as commercial, industrial administrative, transport, religious, medical, recreational and residential.

Gentrification: This is the process whereby city authorities in developed countries are trying to encourage people to return to live in the city usually in redeveloped housing, from former office blocks, which are close to or within the CBD.

Greenbelts: Areas surrounding cities and towns in which laws control development such as housing and industry in order to protect the countryside.

High order functions: These are the most important functions such as department stores, council offices or art galleries.

Inner city: This is the area near the centre which contains basically the CBD, the older manufacturing zone and zone of low cost housing.

Low order functions: These are less important functions such as small corner shops, post offices and petrol stations.

Overspill population: This refers to people who have moved out of the main city to other smaller towns or new towns.

Park and ride schemes: This is an attempt to reduce traffic congestion by encouraging people to park their cars outside city limits and to use their parking ticket to use public transport for the remainder of their journey into the city.

Pedestrianised zones: These are traffic free areas within the city centre where people can shop and walk along streets where traffic is forbidden to enter.

Renewal and regeneration: This refers to the processes by which older areas are demolished and replaced by new buildings, often having totally different functions to the original building or area.

Ring roads: Roads built specifically to take traffic away from the city centre and to help solve the problem of congestion.

Rural-urban fringe: This is the area at the edge of the city where the city limits meet countryside and is attractive to developers for a variety of purposes such as business parks, leisure and recreational parks, out of town shopping centres and motorways. Development may be restricted by legislation such as greenbelt laws.

Shanty towns: These are areas found mainly in and around cities in economically less developed countries and usually consist of temporary and makeshift accommodation made from materials such as wood and corrugated iron. In South American countries they are called 'favelas' and in India they are known as 'bustees'.

Site: This is the actual land upon which a settlement was originally built.

Site factors: These are the factors which influenced people to choose a particular site such as nearness to water supply (river), flat land for building, high land for defence, at a suitable point on a river where a bridge could be built, near raw materials.

Environmental Interactions

Sphere of influence: This is the area from which a settlement draws its customers for various functions. Usually the size of the sphere of influence varies in direct response to the size of the settlement.

Suburbs: Housing zones on the outskirts of towns away from the busy central inner city.

Traffic congestion: Heavy build up of traffic along major routes and within the city centre which causes great problems of cost and pollution for many cities in the UK.

Urban model: This is an idealised view of the internal structure of a settlement and can be used as a basis for comparing settlements.

European Regional Inequalities

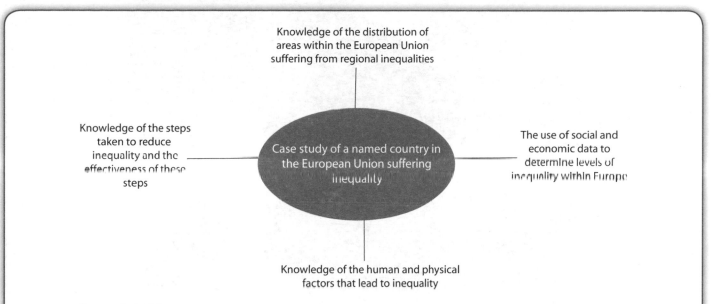

Knowledge of the distribution of areas within the European Union suffering from regional inequalities

Knowledge of the steps taken to reduce inequality and the effectiveness of these steps

Case study of a named country in the European Union suffering inequality

The use of social and economic data to determine levels of inequality within Europe

Knowledge of the human and physical factors that lead to inequality

FIGURE 5.1: MAIN TOPIC AREAS LINKED TO CASE STUDIES OF REGIONAL INEQUALITIES

You can be asked to describe general socio-economic patterns shown on various maps of the EU or countries within it.

Figure 5.2 shows the Euro core and its periphery.

- Areas belonging to the Euro core include, e.g. London and south-east England, Paris and the Ruhr/Rhine area.
- In answers refer to areas by name.
- Name the more well off and poorer areas according to the statistics given.

What socio-economic indicators are used to identify different areas?

These include:

- Average income per head.
- The amount of EU support per head.
- The level of private investment per head.
- The number of cars per head.
- The percentage of long term unemployed.
- The percentage employed in industry.
- The percentage in higher education.
- A combined indicator based on the above.

You can be asked to explain the patterns on maps in terms of physical, social and economic factors.

When answering refer to:

Physical factors

- Climate, which affects tourism, farming and water supply.
- Relief, which affects communications, farming and settlement.
- Landscape features, e.g. mountain ranges/valleys/coastal areas.

Environmental Interactions

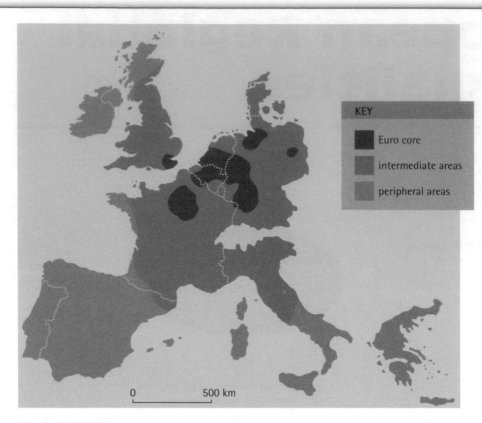

FIGURE 5.2: EURO CORE AND PERIPHERY

Social factors

- Employment rates.
- Life expectancy rates.
- Reasons for these, such as health care, diets, average income and standards of living.
- Literacy rates and the overall standards of education provision.
- Percentage of school leavers in further education.
- Variations in housing standards.

Economic factors

- The level of industrial output.
- Employment data.
- Availability of raw materials.
- The quality of infrastructure.
- GDP or GNP figures.
- Relative percentages employed in industry and agriculture.
- Industrial problems, e.g. declining traditional industries.
- The continuing development of sunrise industries such as electronics.
- The level of government support to industrial areas.

Political factors

- Government intervention.
- Conflict between national and European legislation.
- Political instability.

How do you assess the accuracy of statements on socio-economic patterns?

When doing this mention relevant data which supports your assessment of the accuracy of given statements such as:

- Employment data, e.g. percentages employed in primary, secondary and service industries.
- GDP and GNP figures.
- Migration rates.
- Differences in economic growth rates.

Top Tip

Memorise these for use in answers to questions on this topic.

What problems are faced by peripheral EU areas?

What measures are taken to solve them and how effective are they?

Depending on the areas problems include:

Economic problems

- Low industrial investment.
- A high percentage employed in primary industry.
- A poor infrastructure such as a poorly maintained communications system.
- Long term poverty and low wages.

Social problems

- High levels of unemployment.
- Increasing migration rates, especially emigrants.
- High crime rates.
- High levels of deprivation.
- High welfare dependency.
- Depopulation and ageing populations.

Measures include

- Government funded improvements in education and infrastructure.
- European Union grants and subsidies being made available to poorer areas.
- Poor areas given Development area status.
- Money from the European Development Fund for rural and old industrial areas.

Refer to any statistics which indicate significant changes leading to improved conditions. All these measures have helped to resolve the major social and economic problems.

You can be asked to outline steps taken by the EU and national governments to assist the development of poorer regions, and help with regional inequalities.

When answering, refer to:

- Government incentives.
- Poor areas given regional development status and European Zone status.

Quick Test

List three factors in each of the following categories: Social, Economic and Political.

- Access to government and EU funding.
- Specific assistance is given to areas with traditional industries.
- Governments direct state owned firms to invest in poorer parts of the country.
- Major multinational companies are under pressure to locate in poorer areas.
- Creation of European Regional Development Fund to encourage firms to move to disadvantaged areas.
- Use of European Social Fund which gives grants to improve job opportunities and retraining schemes.
- European Investment Bank to provide loans for factory modernisation.

Topic glossary

Development areas: These are areas which have been designated suitable for financial assistance to help develop and improve agriculture, industry and general infrastructure, e.g. Southern Italy.

Diversification: This is the process whereby an economic enterprise such as a farm takes on a range of additional activities to increase profits, for example, renting land for golf courses, caravan sites, paint ball enterprises, quad biking.

Economic: Relating to financial developments.

Economic effects: The financial impact of change on, for example, employment, incomes, running costs, building costs and costs to the local community.

Enterprise zones: Areas created by governments to stimulate economic development, e.g. Swansea and Milford Haven.

Environmental consequences: The effects of change on the physical and human environments, for example changing land use, improvements and bad effects on the environment, such as increased pollution, changes to the population caused by people moving to or from areas as a result of change, such as industrial closures or new industry being built.

European Development Fund: This is a source of money given to development areas to help them improve their economic condition. It is raised from contributions from member countries of the European Union.

Gross Domestic Product/Gross National Product: This is a measure of the amount of wealth produced in a country in a year.

Industrial decline: This happens when for various reasons industries in an area have to close. This has mainly involved large traditional, heavy industry such as mining, iron and steel making, textiles and shipbuilding.

Industrial growth: This happens when an industry increases its output, number of workers and profits and usually has a positive effect on the local community.

Multiplier effect: This is the wider effect of change on other activities such as other industries, settlements or rural activities. For example, when a major industry closes this may cause local shops and other business to close since people made unemployed have less money to spend and may move elsewhere to seek new work.

Science parks: These are industrial areas which are closely connected to technological institutions and universities and are often involved in producing high technology products.

Development and Health

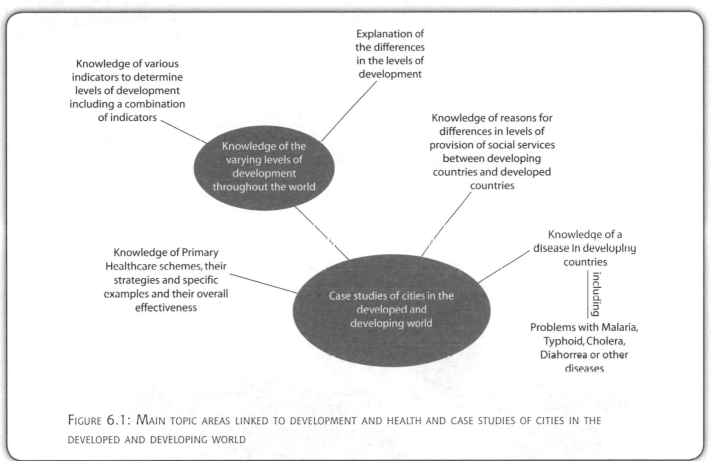

FIGURE 6.1: MAIN TOPIC AREAS LINKED TO DEVELOPMENT AND HEALTH AND CASE STUDIES OF CITIES IN THE DEVELOPED AND DEVELOPING WORLD

Indicators of level of development

What indicators can you use to identify the level of a country's economic and social development?

Economic indicators include:

- Gross National Product or Gross National Product statistics per capita.
- Average income per capita.
- The relative percentage of the workforce employed in industry and agriculture.
- Average electricity consumption kw/capita.
- Import and export figures.
- Trade balances in terms of surplus or deficits.

Social indicators include:

- Birth rates/death rates/infant mortality rates/life expectancy rates per thousand of the population.
- Population structure showing the distribution of age and sex.
- Average calorie intake per capita.
- The average number of people per doctor.
- Literacy rates.

Combined indicators

The use of single indicators can be misleading since the data is based on averages and does not reveal the whole situation. For example, average per capita income or GDP per capita for Saudi Arabia can seem high but income distribution is very unequal. Taking several indicators together produces combined indices such the **Physical Quality of Life Index (PQLI) or the Human Development Index (HDI)**.

This gives a much more accurate picture of the level of development.

In the exam you may be asked to identify the level of development of countries from data such as a table or a map.

Figure 6.2 is an example of this kind of map.

Areas with **low** Gross National Product, **high** birth/death rates, **low** life expectancy, **low** literacy rates, **low** income per capita and a **high** percentage of the workforce working in agriculture would be best described as **Developing** countries.

The opposite of these would help identify **Developed** countries.

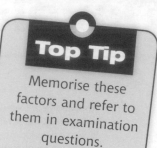

Memorise these factors and refer to them in examination questions.

Make sure that you give named examples of both developing and developed countries in your answers.

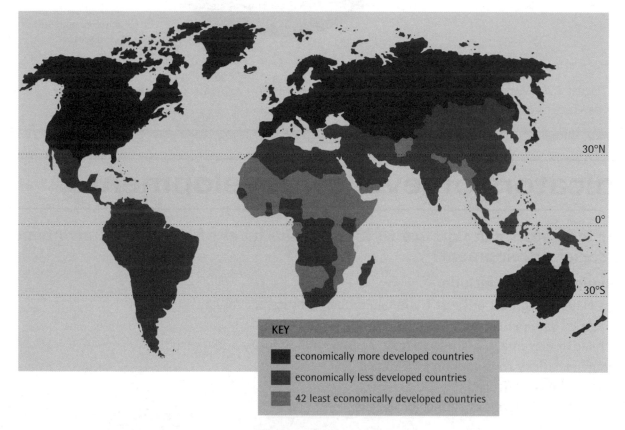

KEY

■ economically more developed countries

■ economically less developed countries

■ 42 least economically developed countries

FIGURE 6.2: LEVELS OF ECONOMIC DEVELOPMENT THROUGHOUT THE WORLD

Quick Test

Make a list of three indicators of development for each of the following headings: Economic, Social and Combined.

You could be asked to describe and account for **similarities and differences** on such maps.

When explaining the factors involved in differing levels of health and disease throughout the world refer to both **physical and human factors** such as:

- **Climate** factors such as rainfall reliability.
- **Economic resources** and agricultural output.
- **Health standards**.
- Presence of **various diseases** such as water borne diseases (cholera, typhoid) and diseases resulting from malnutrition and poor diets.
- Types of **industry**.
- Pattern of **imports and exports**.
- **Demographic trends** with reference to birth, death and growth rates, infant mortality rates and average life expectancy.
- **Average calorie consumption** per person.
- **Housing standards** within cities, noting presence of **shanty towns**.

What factors affect levels of malnutrition?

Refer to the following in answers to this question:

- Diets which are based on one basic food source such as rice lack the variety supplied by nutritious foods.
- This happens in many countries, especially **South East Asia** and the **Indian sub-continent**.
- Consequently much of the population is unhealthy and unable to resist even the most simple of diseases.
- Many people are unable to work for any length of time.
- Industrial and agricultural output suffers.
- Those suffering malnutrition cannot generate enough income for themselves and families to pay for the basic necessities of life.

Do countries have the same provision of safe water and sanitation?

The answer is clearly no. What are the reasons and consequences of this?

Physical factors such as drought affect water supply.

Economic factors, e.g. lack of funding for sanitation and water supply are also responsible.

These problems tend to occur more in poorer agricultural areas which rely on local water sources such as contaminated rivers and streams. Urban areas have sanitation and water supplies but they may be unavailable to people in squatter areas or shanty towns.

Referring to case studies from developing countries explain the impact of lack of clean water and poor sanitation on disease rates.

Depending on the country you have studied, mention:

- Diseases such as cholera and typhoid are transmitted through bacteria carried in dirty water.
- People are forced to cook and drink water from unclean sources.
- Lack of proper sanitation and sewage disposal plants means that raw sewage can infect local water supplies.
- This results in an increased incidence of disease infection.

Top Tip

Make sure you mention at least one country in your answer.

What efforts have been made to improve provision of clean water and sanitation and how successful have they been?

Measures adopted include:

- Producing wells to provide clean water.
- Building water management schemes to provide access to clean water supplies.

- Making improvements in house sanitation and local sewerage plants.
- This has reduced disease rates since disease bacteria which survive in raw sewage are destroyed.
- Many of these efforts are based on foreign aid, charity donations and self help schemes.
- Lack of money and debt problems severely restrict the efforts of the governments of many developing countries.

Top Tip

When revising this topic, go back to the relevant section in the core population section (page 48).

Referring to one disease from a given list, including, for example, malaria, schistomiasis, cholera, kwashiorkor and AIDs, describe the human and physical factors which contribute to the spread of the disease.

Describe methods used to control the disease and to what extent they have been successful.

Malaria is a popular case study.

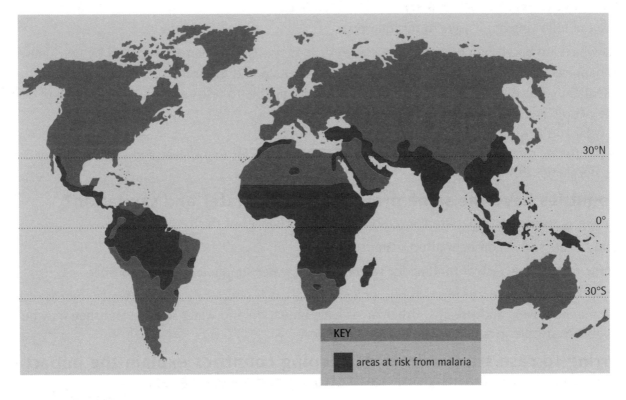

Figure 6.3: Areas at risk from malaria

Quick Test

List three factors which affect each of the following: Birth rates, Death rates, Infant mortality rates and Life expectancy rates.

Malaria

Factors affecting the spread of the disease.

- Malaria is spread by the female anopheles mosquito.
- They take blood meals from infected persons and pass it on through their saliva.
- Mosquitoes breed in stagnant water in hot, wet climates with a minimum temperature of 16 degrees Celsius.
- Without measures taken to limit and control this spread, the disease spreads rapidly.
- Mosquitoes are resistant to many insecticides including DDT.
- The disease has adapted to resist certain drugs formerly used to cure it.
- No vaccine is available to prevent infection.
- Great efforts are currently being made to produce a vaccine.

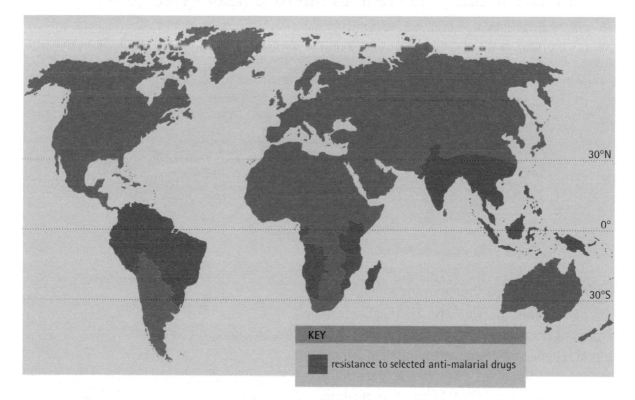

30°N

0°

30°S

KEY

resistance to selected anti-malarial drugs

FIGURE 6.4: RESISTANCE TO SELECTED ANTI-MALARIAL DRUGS

Methods used to control the spread of malaria include:

- Drainage of areas with stagnant water, e.g. swamps to destroy breeding grounds of mosquitoes.
- The use of insecticides such as malathion.
- The use of nets to protect people while sleeping from mosquito bites.
- The use of drugs such as quinine or derivatives of this drug – chloroquin.
- Using village health centres to issue information through primary health care schemes.

Quick Test

If you studied any other disease, from memory make a list similar to the one above of the factors which affect the spread of the disease.

- Releasing water from dams to drown larvae.
- Using egg-white sprayed on to stagnant surfaces to suffocate larvae.
- The introduction of small fish in padi fields to eat larvae.
- The planting of eucalyptus trees to absorb moisture.
- The application of mustard seeds into water areas which drag larvae below water surface and drown them.

Effectiveness of the methods

- These measures have met with varying degrees of success.

- Much depends on local population adhering to suggested precautions and taking regular medication.

- Malaria is still a debilitating and killer disease in many parts of the developing world.

How do primary health care methods improve health standards?

Figure 6.5 shows a summary of primary health care.

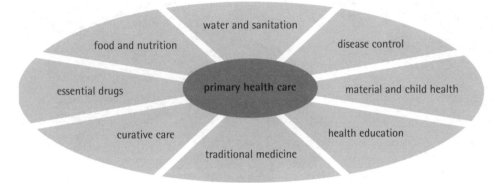

FIGURE 6.5: SUMMARY OF PRIMARY HEALTH CARE

In answering this, mention the following strategies:

- The use of 'barefoot' doctors who are cheaper to train.
- These are not doctors but local people trained in basic first aid medical care.
- Using oral rehydration treatment to tackle diarrhoea.
- Routine medical situations dealt with at local clinics.
- Increased development of a network of local clinics.
- Developing local health education programmes.
- Using cheaper pharmaceuticals rather than more expensive drugs.
- These measures have been very effective especially in reducing costs of medical care.
- Aid programmes which focus on primary health care are often more effective than any which provide large-scale medical programmes.

Quick Test

If you studied a different disease describe how successful the methods used to stop the spread have been.

Sample question, answer and comments

Question

(a) There are often considerable differences in levels of development and living standards within a single country.

Referring to a named ELDC which you have studied, suggest reasons why such regional variations exist.   10 marks

Answer (1) denotes correct point

(a) *In Brazil the north is much less developed than the south. The life expectancy is 63 in the north but 70 in the south. 80% of households in the south have sewage disposal but only 25% in the south.*

The cities of Southern Brazil eg Rio de Janeiro (1) have been able to trade and prosper (1) due to their position on the coasts allowing for ports (1). The soils surrounding cities are suitable for coffee growing which brings a lot of income (1). Rural urban migration caused them to expand increasing the importance of services such as hospitals and schools. The north however is Amazon forest (1) which has poor soils is remote and inaccessible (1) and makes building difficult.

Comments

The question asks for **explanation** for differences within a country. Part of this answer **describes** differences, e.g. the first paragraph, and therefore does not achieve marks for this.

When the answer **explains** differences marks are awarded. There are sufficient explanatory points to obtain a total of **6 marks out of 10**. The important lesson here is to read the question properly and **explain** rather than **describe** as the question asks.

Topic glossary

Appropriate technology: This involves the use of machinery and equipment which is best suited to the needs, skills and wealth of local communities.

Bilateral aid: This is aid given from one country to another.

High technology: This is the use of advanced, sophisticated machinery which requires a high degree of skill to operate.

Intermediate technology: This is machinery and equipment which is of a higher level than basic primitive equipment but not as advanced as high technology.

Long term aid: This is aid which is intended to be used over a long period of years to help a country to develop, for example, its industry, farming, transport system, education and health care systems.

Low technology: This is equipment which is very basic and cheap such as ox-drawn ploughs, wood burning ovens.

Multilateral aid: This is aid given from a group of countries through agencies such as United Nations to poorer countries.

Official aid: This is either bilateral or multilateral aid given to a country.

Primary health care: This is a system designed to provide basic health and medical care to people in economically less developed countries which is cost effective and more readily available to people suffering from relatively minor health complaints. Rather than use highly trained medical staff or expensive hospitals, it relies on people who have basic medical skills and is therefore available to a larger number of the population.

Self-help schemes: These are projects in which local people become involved to improve their living conditions.

Short term aid: This is aid given immediately to help an area recover from a major disaster such as a flood, drought, famine or earthquake.

Tied aid: This is aid given but with conditions; usually dependent on the receiving country using the money given to buy manufactured goods from the donor country.

United Nations Organisation: A worldwide organisation of which nearly all of the world's countries are members. It has a wide range of functions such as world health, world finance, peacekeeping forces and various aid agencies.

Voluntary aid: This is aid given through charitable organisations such as Red Cross or Oxfam or Save the Children Fund.